Information and Measurement

Second Edition

RENEWALS 458-4574

WITHDRAWN
UTSA LIBRARIES

Information and Measurement

Second Edition

J. C. G. Lesurf

Physics and Astronomy Dept
University of St Andrews, Scotland

I$_o$**P**

Institute of Physics Publishing
Bristol and Philadelphia

Library
University of Texas
at San Antonio

© I. O. P. Publishing Ltd, 2002

All rights reserved. No part of this publication may be reproduced, stored in a retrieval system or transmitted in any form or by any means, electronic, mechanical, photocopying, recording or otherwise, without the prior premission of the publisher. Multiple copying is permitted in accordance with the terms of licences issued by the Copyright Licensing Agency under the terms of its agreement with the Committee of Vice-Chancellors and Principals.

British Library Cataloguing-in-Publication Data

A catalogue record for this book is available from the British Library.

ISBN 0 7503 0823 0

Library of Congress Cataloging-in-Publication Data available

First published 1995 (Hardback)
Revised and extended second edition published 2002 (Paperback)

The right of James Lesurf to be identified as the author of this work has been asserted by him in accordance with the Copyrights, Designs, and Patents Act, 1988.

Published by Institute of Physics Publishing, wholly owned by
The Institute of Physics, London

Institute of Physics Publishing, Dirac House, Temple Back, Bristol
BS1 6BE, UK

US Office: Institute of Physics Publishing, The Public Ledger Building, Suite 1035, 150 South Independence Mall West, Philadelphia,
PA 19106, USA
Printed in the UK by J. W. Arrowsmith Ltd, Bristol.

Library
University of Texas
at San Antonio

To Chris, as always, for making everything worthwhile.

Special thanks to Mike Glover and Bob Pollard
at Icon Technology for *TechWriter*, the
technical desktop publisher which
made writing this book a pleasure!

Contents

Contents

Contents

Contents

Preface to the 2nd Edition

This new edition contains over 50 pages of new material. Most of this is contained in three entirely new chapters. These deal with counting, frequency measurement, and the use of correlation to detect and identify signal patterns. In addition, the original version of Chapter 19 on *Data Thinning* has been replaced by an entirely new chapter. The version in first edition was relatively brief and chose the ill-fated example of DCC (Digital Compact Cassette). The new version in this book is substantially longer and uses JPEG and MiniDisc as its examples. As well as these major changes, the opportunity has been taken to correct some minor errors and omissions.

The flavour and intent of the book remains unchanged, but I hope that the changes will enhance the book's usefulness. As before, the approach I have taken is to base explanations upon the underlying physics and use examples which the reader may be familiar with and find interesting.

Jim Lesurf
February 2001

Preface to the 1st Edition

Information has many faces. A physicist may take a course called *Instrumentation* or *Measurement Techniques*. An engineer may study *Information Technology*, and a computer scientist or mathematician *Information Theory*. Courses under these and similar names all tend to offer partial views of a bigger underlying subject.

The specialisation of students taking different degree subjects has tended to lead to a visible fragmentation in the coverage of existing textbooks. On the one hand there are many theoretical books dealing with the mathematics of information theory which ignore the engineering required to put theory into practice. On the other hand there are engineering books on instrumentation technology which fail to give a clear explanation of the concepts which underpin their operation. However, to collect information we have to make measurements. We need real, practical instruments to collect and process this information. A pattern of numbers on a computer disc or a waveform on an oscilloscope screen tells us nothing unless we know how it was produced.

The main purpose of this book is to provide a <u>readable</u> and <u>interesting</u> introduction to a subject area wide enough to be useful to almost every scientist and engineer. The emphasis is on width and clarity rather than an attempt to include every detail. In my experience many undergraduate students find information theory textbooks too abstract and mathematical. This tends to deter all but the most theoretically minded from understanding the subject. Yet information technology is arguably the most important scientific topic of all for anyone who wants to understand and participate in the new technologies which dominate our society. To be useful, the mathematics of information theory has to be based on the properties of the real world and lead to practical applications. As a result, the apparently distinct topics often called information theory, measurement, and instrumentation are best understood by recognising that they are facets of the same jewel.

The approach I have taken in this book differs from most other texts. I have deliberately mixed together the basic maths, engineering, and physics in order to show how they are linked in real situations. I have chosen to illustrate the basic techniques of measurement and information processing using examples which are likely to be of interest to most science and engineering students (and, I suspect, their teachers!). For this reason a large portion of the book concentrates on the *Compact Disc* audio system. There are also chapters on *Encryption* (secret codes) as well as chaos and its uses. The CD system is particularly useful — both because most of us will have encountered it, and because it provides an excellent illustration of how measurement and information technology go together in the real world. The other examples show the range and power of the subject.

For engineers and scientists 'absolute truth' is a matter of personal judgement, not objective fact. In information theory this means that every measurement and message only conveys a finite amount of information. In the real world nothing is absolutely certain or precise. Our state of knowledge is always imperfect, limited, and subject to later improvement.

Preface

I have tried to to avoid the error — sadly common in textbooks — of presenting every detail and ramification of an argument and burying understanding under a mound of facts. This book explains the concepts of information theory on a 'successive approximation' basis. The explanations given in each chapter are intended to be simple enough to guide the reader through the subject without causing confusion. Later explanations give further details as required when more sophisticated techniques are introduced.

If the book has a theme it is that 'The best place to start is the physics of the real world's behaviour'. The form of the book is designed to make it suitable as a 'course book' for an undergraduate course of up to a couple of dozen one-hour lectures. Each chapter provides the material for one lecture topic. Each finishes with a summary which the reader can use to check that they have learned the main points. Most chapters are also followed by a set of tutorial questions. Detailed answers to the numerical questions are provided in an appendix. The correct answer value is also included (in bold type) at the end of each numerical question. You can use this to check your answer before consulting the back of the book. There is an additional appendix listing a number of programs in both BASIC and 'C'. The purpose of these questions and programs is to help the reader to discover how the ideas presented in the book are put into practice.

I hope that I have produced a book which will be useful to a wide range of physical scientists, engineers, mathematicians, and computer scientists. If I have been successful this book will help illuminate how their individual interests and skills link together to form a greater body of understanding. Finally, I would like to thank all the students and others who helped me to discover and correct the mistakes which earlier versions of this book contained. They provided a powerful error correction mechanism I haven't described in the book!

Jim Lesurf
July 1994

Chapter 1

Where does information come from?

1.1 Introduction

This book is designed to provide you with an explanation of the basic concepts of information collecting and processing systems. To do this we will examine examples ranging from secret codes to compact disc players. Using these practical examples you should be able to see how the mathematics of *Information Theory* can be applied in practical situations to make *Instruments* which perform useful tasks. This first chapter is intended to a be a general outline. Most of the concepts introduced here will be looked at more carefully later.

Scientists and engineers devote considerable attention to the processing and storage of information, yet questions relating to how information is produced generally attract less consideration. To some extent, this blind spot seems to stem from a belief that any interest in this area smells strongly of philosophy, not engineering. In general, practically minded scientists don't want to 'waste their time' with philosophy — although there are many notable exceptions to this rule.

This book is <u>not</u> about philosophy. No time will be devoted to questions like:
'What is the meaning of meaning?'
'How do we know what we know?'
 etc.
Despite this, when trying to understand information based systems it's vital to have some idea of how information is created or captured.

*1.2 What **is** information?*

For our purposes, we can say that information initially comes from some form of <u>sensor</u> or transducer. This generates some form of response which can then be measured. It is this measurement or detection which 'creates' information. (In fact, the sensor is reacting to the arrival of some input pattern of energy or power. It would be fairer to say it 'picks up' the

1

information, but we'll ignore this fact.) Once we adopt this starting point it becomes clear that the topics of instrumentation and measurement form the basis of all practical information systems.

This viewpoint provides us with a double advantage over someone who is studying information theory purely as a branch of mathematics. Firstly, it gives us a way to understand information processing systems in terms of the physical properties of the real world. Secondly, it helps us sort out questions related to the 'value' or 'meaning' of information without the risk of being dragged into metaphysics. Instead we can simply ask, 'How was this information produced?'

What is an 'instrument'? At first glance, it can appear to science and engineering students that the subject called *Instrumentation* is obsessed with describing how voltmeters and oscilloscopes work. Yet the subject covers a much wider and more important area. A colour TV is an instrument. A digital computer is an instrument. Each senses some form of input and responds by producing an appropriate output. The TV responds to an electromagnetic wave from a distant transmitter to produce a corresponding picture on a screen and sound from a loudspeaker. The computer can be affected by various sorts of input, from a keyboard, a mouse, or by reading a magnetic disc. It can respond by altering the electronic pattern held in its memory, by altering its monitor display, or recording something on a disc.

Most of the examples we'll look at in this book will be electronic or optical. This is because optical and electronic methods are powerful and widely used. Despite this, it's important to realise that the basic points made in this book aren't only true in these areas. To emphasise this, we can start by considering a simple mechanical measurement system — a kitchen balance — to make some fundamental points which apply to all measurement (information gathering) systems.

The balance has a pan or plate supported by a spring. When we place something on the pan the added weight presses down on the spring, compressing it. The pan moves downwards until the compression force from the squeezed spring balances the force of the increased weight. Most balances have a rotary dial with a pointer attached to the pan. The movement caused by the weight rotates the pointer to give us a 'reading' of the weight.

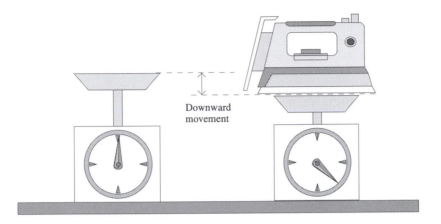

Figure 1.1 Kitchen technology measurement system.

The first point to note is that, like most measurement systems, this one is *indirect*. What we actually observe is a movement (rotation) of the pointer. We don't actually see the magnitude of the weight. If, for example, we put an iron on the pan we might see the pointer move around through 120 degrees. If we liked, we could also use a ruler to find that the pan moved down 2 cm. However, we don't usually quote weights in degrees or centimetres! In order to make sense of these observed values we have to *calibrate* the balance. To do this we can place two or three different <u>known</u> weights on the scales and make a note of how far the pointer goes around (or the pan falls) each time. We can then use these results to make a series of calibration marks on the face of the dial. Now, when we put something — e.g an iron — on the scales we can read off its weight from the dial. This calibration process means that the balance provides us with a means to compare the weight of the iron with a set of other 'standard' weights. In general, <u>all</u> measurements are *Comparisons* with some defined standard.

Usually, we buy a kitchen balance which should already be calibrated (i.e. its dial is marked in kg, lb, etc, not degrees) and we don't bother to calibrate the weighing instrument for ourselves. However, when we consider the need for a calibration process an awkward question springs to mind — where did the 'known' weights come from that were used to calibrate the readings? If all measurements are comparisons, how were the values of <u>those</u> weights known? They, too, would need to have been weighed on some weight measurement system. If so, how was <u>that</u> system calibrated?

Any measurement we make is the last link in a chain of similar measurements. Each one calibrates a system or a 'standard' (e.g. a known weight) which can be used for the next step. Right back at the beginning of this chain (at the National Physical Laboratory in the UK and other standards labs around the world) there will a *Primary Reference* system or standard which is used to <u>define</u> what we mean by '1 kg', or '1 second' or whatever. In effect, when we plonk something on the pan of a kitchen balance we're indirectly comparing it with the standard kg weight kept under a glass cover at the NPL.

When we place an iron on the pan we have to wait a second or two to let the system settle down and allow the pointer to stop moving. Similarly, when we remove the iron the system takes a short time to recover. The second point we can make about the measurement system is, therefore, that it has a finite *Response Time* — i.e. we have to wait for a specific time after any change in the weight before we can make a reliable reading. This limits our ability to measure any changes which take place too quickly for the system.

The third point to note is fairly obvious from our choice of an iron. If we put too large a weight on the pan the pointer will go right around and move 'off scale'. (If the iron is very heavy we may even smash the scales!) No matter how well we search the shops, we can't find scales which can accurately measure <u>any</u> weight, no matter how big. Every real instrument is limited to operate over some finite *Range*. Beyond this range it won't work properly and *Overloads* or *Saturates* to give a meaningless response.

The fourth and final basic point is something we won't usually notice using ordinary scales since the effect is relatively small. All of the atoms in the scales, including those in its spring, will be at room temperature. (In a kitchen this probably means at or above 20 Celsius or 293 Kelvin.) As a result, they'll be moving around with random thermal motions. Compared to the effect of placing an iron on the scales these movements are quite small. However, if we looked at the pointer very carefully with a powerful microscope we'd see its angle fluctuating randomly up and down a little bit because of the motions of the atoms in the spring. As a result, if we wanted to measure the weight very accurately this thermal jittering would limit the precision of our reading. As a result, no matter how good the scales, our ability to make extremely accurate measurements is limited by thermal random effects or thermal *noise*.

4

1.3 Accuracy and resolution

It is important to realise that the amount of information we can collect is always finite. The example of kitchen scales has introduced us to the limiting effects of clipping, noise, and response time. It doesn't matter how clever we are, these problems occur in all physical systems since they are consequences of the way the real world works. To see some of the other problems which arise when we're collecting information, consider the system in figure 1.2. This diagram represents a diffraction grating being used to measure the power/frequency spectrum produced by a light source.

The system is intended to provide us with information about how bright the light source is at various light wavelengths. It relies upon the reflection properties of a surface made with a series of parallel ridges called a *Reflection Grating*. For an ordinary plane mirror, the angle of reflection equals the angle of incidence. For a grating, the angle of reflection also depends upon the wavelength of the light and the details of the grooved surface pattern. Hence the arrangement shown acts as a sort of adjustable filter. Only those light wavelengths which reflect at the appropriate angle will make their way through the output slit onto the detector.

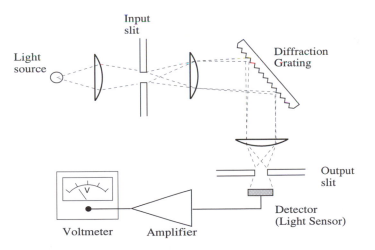

Figure 1.2   Simple diffraction grating spectrometer.

As with the kitchen scales, the system provides an indirect way to measure the light's spectrum. We use the angle of the diffraction grating to tell us the wavelength being observed. The voltage displayed on the meter indicates the light power falling on the detector. To discover the light's spectrum we slowly rotate the grating (or move the output lens/slit/ detector) and note how the voltmeter reading varies with the grating angle. To convert these angles and voltages into wavelengths and light powers we then need to know the *Sensitivity* of the detector/amplifier system and the angles at which various wavelengths would be reflected by the grating – i.e. the system must be calibrated.

In most cases the instrument will be supplied with appropriate display scales. The voltmeter will have a dial marked in units of light power, not volts. The grating angle display will be marked in wavelengths, not degrees. These scales will have been produced by a calibration process. If the measurements we're making are important it will probably be sensible to check the calibration by making some measurements of our own on a 'known' light source.

As with the kitchen balance, our ability to measure small changes in the light level will be limited by random noise — in this case random movements of the electrons in the measurement system and fluctuations in the rate at which photons strike the detector. The accuracy of the power measurement will depend upon the ratio of the light power level hitting the detector to the random noise. We could increase the light level and improve the precision of the power measurement by widening the slits and allowing more light through. However, this would have the disadvantage of allowing light reflected over a wider *range* of angles to reach the detector. Since the angle of reflection depends upon the light wavelength this means we are allowing through a wider range of wavelengths.

In fact, looking at the system we can see that it <u>always</u> allows through a range of wavelengths. Unless the slits are narrowed down to nothing (cutting off all the light!) it will always allow light reflected over some range of angles, $\Delta\Theta$ (and hence having a range of wavelengths, $\Delta\lambda$) to get through. As a result there is an unavoidable 'trade off' between the instrument's power sensitivity and its frequency *Resolution* or ability to distinguish variations in power confined to a narrow frequency interval. This kind of trade off is very common in information collection systems. It stems from basic properties of the physical world and means that the amount of information we can collect is always finite — i.e. we can <u>never</u> make perfect measurements with absolute accuracy or precision or certainty.

Summary

You should now know that information is collected by *Instruments* which perform some kind of *Measurement.* That measurement systems usually give an *Indirect* indication of the measured quantity and that all measurements are *Comparisons* which have to be *Calibrated* in some way. The amount of information we can collect is always finite, limited by the effects of *Noise, Saturation* (or *Overload*), and *Response Time.* That many information gathering techniques involve a *Trade-Off* between various quantities — for example, between the *Resolution* of a wavelength measurement and the *Sensitivity* of a related power measurement. That these limitations arise from the properties of the physical world, <u>not</u> poor instrument design.

Chapter 2

Signals and messages

2.1 Sending information

In the first chapter we looked at how measurement instruments can produce information. The information produced and processed by these systems will be in the form of a *Signal* which carries a *message* (specific set of information). In the case of the grating spectrometer shown in figure 1.2 the signal was a voltage communicated from the light detector to the voltmeter. This voltage will vary in a specific pattern as the grating angle is altered. It is this pattern which carries the message.

All information handling systems have the same basic form. Firstly, there will be some type of information *Source*. This can take many forms, from the microphone in a telephone to the keyboard of a computer. The source will be connected to a *Receiver* by some sort of *Channel*. In the case of a telephone, the receiver will be an earpiece in another telephone and the information carrying channel between them may be a set of wires. Information is sent along the wires in the form of a varying voltage and current which acts as a signal whose details carry the actual information or message.

In this book we will tend to talk about signals being 'transmitted'. Despite this it's important to realise that — from the theoretical point of view — there isn't any real difference between transmitting signals, storing them on discs/tapes etc to read later, and processing them in a computer. Most of the basic comments and properties outlined in this book apply to information processing systems in general. They aren't restricted to telephones or TV broadcasts! For this reason the concept of signals is of fundamental importance to information theory. Before the invention of the telephone, people could send messages by posting written letters, or by getting a chain of other people to stand on hilltops and wave semaphore flags, or even by lighting bonfires! Before the desktop computer there was pen and paper. Modern systems are more convenient, but if you really wanted to you could do it some other old-fashioned way!

No matter how it's done, before a signal can be used to communicate some specific information in the form of a *Message*, the sender and receiver must have agreed on the details of how the actual signals are to be used. It is not enough to agree that someone will stand on a hilltop and wave flags. We have to arrange that, "These flags held like this represent the letter 'A'; these held like this represent 'B'..." i.e there must be some sort of pre-arranged *Code* for sending the information. It is also clearly important that we can distinguish one code *Symbol* ('A', 'B', 'C', etc are examples of distinct symbols) from another, otherwise we will make mistakes.

On the basis of 'A=000, B=001, C=010, etc...' this signal could be sampled at the points shown and then sent in the form, 'GGGHHFED...', etc.

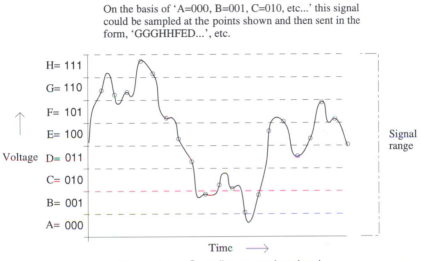

Figure 2.1   Sampling an analog signal.

It's important to realise that the same message can be conveyed in any form we like provided we obey the basic rules of information theory. As an example, consider the message illustrated in figure 2.1. This shows a varying voltage coming from a sensor. At this point it doesn't matter very much where this pattern has come from or what it represents. It might be coming from a telephone mouthpiece and carrying information about what someone is saying. It might be from the light detector in figure 1.2 and indicates how the light level varies as the grating angle is altered. What matters is that the details of the signal pattern constitute the message which carries the information. In the case of the instrument shown in figure 1.2 the information is signalled from detector to voltmeter by a smoothly varying voltage whose level is roughly proportional to the detected light level. Signals of this type are called *Analog* since the varying level (the voltage) is treated as a mathematical analog of the original (light power in this case) pattern. We can therefore imagine that the

9

shape of the curve plotted in figure 2.1 holds the information about the spectrum of the light being observed.

If we wanted to communicate this information to someone we could connect up some amplifiers and wires and send it as an analog voltage level which varies as shown. (In this case, the various voltage levels which we can distinguish from one another are the 'symbols', although it's not normal for analog signals to be described in that way.) Alternatively, we can adopt other ways to communicate or store the same information. For example, we can choose to *Sample* the signal waveform and convert it into a series of binary numbers. To do this we proceed as follows.

We begin by defining a specific maximum *Signal Range* which is wide enough to ensure that the signal level is always inside the chosen range. We then choose a point on the waveform and ask, 'Is the point in the top half of the range?'. If it is we write down a '1', if not we write down a '0'. We then define a <u>new</u> range which only covers that half of the original one which contains the point and ask the question again to obtain another '1' or '0' answer. This provides a two-digit number which tells us which quarter of the original range the point occupies. In principle, this process of halving the range, asking the question, getting a yes/no answer, and noting the result as a one or zero can be repeated as many times as we like. We can then repeat this whole process for a series of points along the waveform. This process is called *Sampling* the waveform. Note that if the signal level ever moves out of the initial signal range we've chosen we won't have any way of indicating its actual level. Should this happen, the signal is said to have been *Clipped* since we can only indicate its value by the nearest available set of '1's and '0's.

In the example shown in figure 2.1 the question and answer process is performed three times for each chosen point. This gives us a series of three-digit values which tell us which eighth of the signal range contains each sample. The result is a series of binary numbers whose pattern holds the information required to define or reconstruct the actual waveform. We could therefore transmit these numbers to someone and they could then use them to draw out the original waveform shape.

The process considered above converts the waveform information into a signal encoded in *Binary Digital* form. Digital numbers are very convenient to transmit and are ideal for storing and processing in modern digital computers. We can, however, encode the same information in any way we find convenient. For example, if we wanted to record in a notebook, we

10

could represent each possible digital number as a letter. For example, as shown in the diagram, we could choose 000 = 'A', 001 = 'B', 010 = 'C', etc. The information in the waveform could then be written down as 'GGGHHFED...'. It doesn't matter what form of code we choose. Provided we have *Encoded* it correctly, the same information will be preserved. The message will remain the same although the form of the signal used (analog voltage, digital numbers, letters in a book) will be different.

2.2 How much information in a message?

In the above example we asked three yes/no questions about each chosen point on the initial waveform. Yes/no questions like this are the simplest we can ask. Each answer is a yes/no or '1'/'0' which gives us the minimum possible amount of extra information. This minimum possible quantity of information is called a *Bit*. Having asked three yes/no questions per point we therefore obtain a series of values, each of which contains just three bits worth of information. In general, asking n questions per sample produces a series of n-bit binary numbers, each of which defines which $1/2^n$th of the signal range each point occupies. There are only 2^n possible n-bit numbers. Hence we require 2^n distinct symbols ('A', 'B', 'C', ... 'H' or '000', '001', '010', ... '111', or whatever) to convey the information. The limited range of possible values means we can use a limited 'alphabet' of 2^n symbols.

The amount of information we collect about the waveform depends upon how many points we sample and how many yes/no answers we get for each. We can therefore hope to get twice as much information by taking double the number of samples. However, although asking an extra question per sample doubles the number of symbols required it <u>doesn't</u> provide twice as much information. In the example considered above, asking an extra question per sample would mean each binary result would have four bits instead of three. This means we would collect four-thirds as much information <u>not</u> twice as much! The basic rule of information theory is that the total amount of information, H, collected will be

$$H = Nn \qquad \qquad ...(2.1)$$

where N is the number of samples and n the number of bits (questions and answers) per sample. Given an initial signal which lasts for a period of time, T, sampled at a series of instant t apart, we would therefore obtain a total amount of sampled information

$$H = \frac{Tn}{t} = \left(\frac{T}{t}\right) \log_2\{M\} \qquad \qquad ...(2.2)$$

11

where $M = 2^n$ is the number of <u>symbols</u> available to convey the message.

In practice, the amount of information we can communicate in a given time will be limited by the properties of the *channel* (the wires, amplifiers, optical fibres, etc) we use. We therefore often need to know the information carrying *capacity* of a channel to decide if it is up to a given task. Consider the example of a varying analog voltage sent along some wires to be measured with an *Analog to Digital Convertor* (ADC). Here the wires are the channel and the ADC is the signal receiver. How many bits worth of information could an ideal ADC obtain from the analog signal in a given time? The input seen by the ADC will be a combination of the transmitted signal level and a small amount of random noise. This determines the size of the smallest signal details we can expect to observe. There will also be a limit to how great a signal voltage can be transmitted without 'clipping' or serious distortion. For the sake of example, let's assume that the channel has a noise level of around 1 mV and can handle a maximum range of 1 V.

Ideally, the ADC's range should equal that of the input channel, i.e. the ADC should in this case start with a voltage range, V_{range} of 1 V. An *n*-bit ADC could then determine the signal level at any instant with an accuracy of $V_{range} / 2^n$. An 8-bit ADC could divide the input 1 V range into $2^8 = 256$ bands, each $1 / 2^8 = 0 \cdot 0039$ V wide, and determine which of these bands the input was in at any instant. A 10-bit ADC could divide the 1 V range into $2^{10} = 1024$ bands, each $1 / 2^{10} = 0 \cdot 00097$ V wide. We might therefore expect to extract more information and obtain a more accurate result by using a 10-bit ADC instead of an 8-bit one. However, if we tried using an even better ADC giving 11 or more bits per sample we <u>wouldn't</u> obtain any extra information about the signal. This is because the 10-bit ADC already divides the input range into bands just $0 \cdot 97$ mV wide — i.e. slightly smaller than the amount by which the random noise jitters the input up and down. There's no point trying to determine the voltage level more accurately than this. We'll simply be looking at the effects of the noise. So it would a waste of effort to use an 11-bit ADC in this case as the 'extra' bits wouldn't tell us anything useful.

This effect arises because the input signal has a finite *Dynamic Range* — the ratio of maximum possible signal size to the minimum detail detectable over the random noise. The dynamic range, *D*, of an analog signal is defined as a <u>power</u> ratio given in decibels between the maximum possible signal level and the mean noise level, i.e. we can say that

$$D = 10 \log\left\{\frac{P_{max}}{P_n}\right\} = 10 \log\left\{\frac{V_{max}^2}{v_n^2}\right\} \qquad \text{...(2.3)}$$

where V_{max} and v_n represent the rms maximum signal and rms noise. This dynamic range should be distinguished from the actual *signal to noise ratio* (SNR), at any time

$$\text{SNR} = 10 \log\left\{\frac{P_s}{P_n}\right\} \qquad \text{...(2.4)}$$

where P_s is the actual signal power level which is usually less than P_{max}.

There will also be a limitation on how quickly the voltage being transmitted along the wires can be changed. This is due to the finite response time of any system. Here, for example, we can assume that (due, perhaps to stray capacitances) the wires take a microsecond to react to a change. This means we can't expect to obtain any extra information by making the ADC sample the input it sees more often than once a microsecond, choosing a *sampling rate* above 1 MHz (10^6 samples per second) won't therefore provide any extra information.

If it takes the channel (the wires) a microsecond to respond to a voltage rise and a microsecond to respond to a fall, the highest signal frequency we can expect it to carry will be one cycle (one up and down) every <u>two</u> microseconds — a maximum signal frequency of 0·5 MHz. The *Bandwidth* of a channel is the range of frequencies it can carry. In most cases we can assume that this range extends down to 'd.c.' so the maximum frequency and the bandwidth usually have the same value. In this case we see that the sensible sampling rate is about 1 MHz and the bandwidth of the analog channel is 0·5MHz. This implies that, in general, we can expect the required sampling rate to be double the bandwidth.

In this case, the combination of 1 mV of noise, a signal voltage range of 1 V, and a 1 µS response time mean that there is no point in using an ADC which tries to collect more than 10 bits per microsecond. It is important to note that this limitation of the rate the ADC collects information is imposed by the <u>channel</u> which transmits the analog signal to it, <u>not</u> a defect of the ADC itself. A better ADC wouldn't give us any extra information since 10×10^6 bits per second is all this particular analog signal channel can carry. The analysis we've carried out here is just a rough approximation. We'll be considering the question of the information carrying *capacity* of a channel more carefully in a later chapter. However, we can already see that the effects of random noise, clipping, and response time/bandwidth combine to limit the information carrying

13

capacity of <u>any</u> information channel no matter what form of signal it uses.

Summary

In this chapter you saw how all information processing systems can be regarded as consisting of an information *Source* connected to a *Receiver* by some form of *Channel*. That any particular set of information is a *Message* which is sent as a *Signal* pattern using some form of *Code* made up of appropriate *Symbols*. You saw how an analog (continuously varying) signal can be *Sampled* to recover <u>all</u> the information it contains. That the amount of information a channel carrying an analog signal can convey is finite, limited by the biggest unclipped level it can manage (*Clipping*), the *Noise* level, and the time it takes to respond to a changed input (the channel's *Response Time* or *Bandwidth*).

Questions

1) Sketch a diagram of a typical analog *Signal* pattern. Use the diagram to help explain how such a signal can be *Sampled*, and what we mean by a *Bit* of information.

2) An analog voltage *Channel* is used to transmit a signal to an *Analog to Digital Converter* (*ADC*). The input voltage can vary over the range from +2 to −2 V and the channel *Noise* level corresponds to ±1 mV. How many bits per sample must the ADC produce to be able to measure the input voltage level at any moment without any loss of information? How many different code *Symbols* would be required to record all the possible values produced by the ADC? [**11 bits/sample. Minimum of 2000 symbols needed to cover all the levels. The 11-bit ADC actually provides 2048 symbols.**]

3) The channel used for 2) can carry signal frequencies (sinewaves) from 0 Hz up to 150 kHz. What is the value of the channel's *Response Time*? How many samples per second must the ADC take to ensure that all the analog information is converted into digital form? [**Response time = 3·3 µs. 300,000 samples/s.**]

4) A *Message* takes 10 seconds to transmit along the analog channel. How many bits of of information is it likely to contain? [**33 million.**]

5) Explain the difference between the *Dynamic Range* of a channel or system and the *Signal to Noise Ratio* of a signal. Write down an equation giving the S/N ratio in decibels in terms of the signal power and noise power.

Chapter 3

Noise

3.1 The sources of noise

Whenever we try to make accurate measurements we discover that the quantities we are observing appear to fluctuate randomly by a small amount. This limits our ability to make quick, accurate measurements and ensures that the amount of information we can collect or communicate is always finite. These random fluctuations are called *Noise*. They arise because the real world behaves in a quantised or 'lumpy' fashion. A common question when designing or using information systems is, 'Can we do any better?' In some cases it's possible to improve a system by choosing a better design or using it in a different way. In other cases we're up against fundamental limits set by unavoidable noise effects. To decide whether it is worth trying to build a better system we need to understand how noise arises and behaves. Here we will concentrate on electronic examples. However, you should bear in mind that similar results arise when we consider information carried in other ways (e.g. by photons in optonics systems).

3.2 'Johnson noise'

In 1927 J. B. Johnson observed random fluctuations in the voltages across electrical resistors. A year later H. Nyquist published a theoretical analysis of this noise which is thermal in origin. Hence this type of noise is variously called *Johnson* noise, *Nyquist* noise, or *Thermal* noise.

A resistor consists of a piece of conductive material with two electrical contacts. In order to conduct electricity the material must contain some charges which are free to move. We can therefore treat it as 'box' of material which contains some mobile electrons (charges) which move around, interacting with each other and with the atoms of the material. At any non-zero temperature we can think of the moving charges as a sort of *Electron Gas* trapped inside the resistor box. The electrons move about in a randomised way — similar to Brownian motion — bouncing and scattering off one another and the atoms. At any particular instant there may be more electrons near one end of the box than the other. This

16

means there will be a difference in electric potential between the ends of the box (i.e. the non-uniform charge distribution produces a voltage across the resistor). As the distribution fluctuates from instant to instant the resulting voltage will also vary unpredictably.

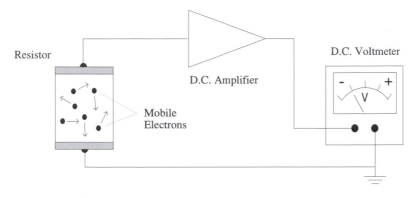

Figure 3.1 Fluctuating voltage produced by random movements of mobile electrons.

Figure 3.1 illustrates a resistor connected connected via an amplifier to a centre-zero d.c. voltmeter. Provided that the gain of the amplifier and the sensitivity of the meter are large enough we will see the meter reading alter randomly from moment to moment in response to the thermal movements of the charges within the resistor. We can't predict what the precise noise voltage will be at any future moment. We can however make some <u>statistical</u> predictions after observing the fluctuations over a period of time. If we note the meter reading at regular intervals (e.g. every second) for a long period we can plot a histogram of the results. To do this we choose a 'bin width', dV, and divide up the range of possible voltages into small 'bins' of this size. We then count up how often the measured voltage was in each bin, divide those counts by the <u>total</u> number of measurements, and plot a histogram of the form shown in figure 3.2.

We can now use this plot to indicate the likelihood or *probability*, $p\{V\}.dV$, that any future measurement of the voltage will give a result in any particular small range, $V \rightarrow V + dV$. This type of histogram is therefore called a display of the *Probability Density Distribution* of the fluctuations. From the form of the results, two conclusions become apparent:

Firstly, the <u>average</u> of all the voltage measurements will be around zero

17

volts. This isn't a surprise since there's no reason for the electrons to prefer to concentrate at one end of the resistor. For this reason, the average voltage won't tell us anything about how large the noise fluctuations are.

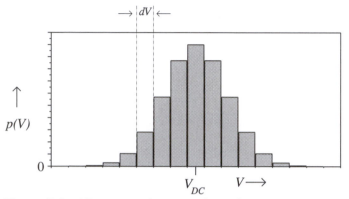

Figure 3.2   Histogram of some noise voltage measurements.

Secondly, the histogram will approximately fit what's called a *Normal* (or *Gaussian*) distribution of the form

$$p\{V\} \cdot dV \propto \text{Exp}\left\{\frac{-2V^2}{\sigma^2}\right\} \qquad \qquad \text{...(3.1)}$$

(Note that you'll only get these results if you make <u>lots</u> of readings. One or two measurements won't show a nice Gaussian plot with its centre at zero!) The value of σ which fits the observed distribution indicates how wide the distribution is, hence it's a useful measure of the amount of noise.

The σ value is useful for theoretical reasons since the probability distribution is Gaussian. In practice, however, it is more common to specify a noise level in terms of an *rms* or *root-mean-square* quantity. Here we can imagine making a series of m voltage measurements, v_1, v_2, ... v_j ... v_m, of the fluctuating voltage. We can then calculate the rms voltage level which can be defined as

$$v_{rms} \equiv \sqrt{\sum_{j=1}^{m} \frac{v_j^2}{m}} \qquad \qquad \text{...(3.2)}$$

In general in this book we can simplify things by using the 'angle brackets', $\langle \rangle$, to indicate an averaged quantity. Using this notation expression 3.2 becomes

18

$$v_{rms} = \sqrt{\langle v_j^2 \rangle} \qquad \ldots (3.3)$$

Since v_j^2 will be positive when $v_j > 0$ <u>and</u> when $v_j < 0$ we can expect v_{rms} to always be positive whenever the Gaussian noise distribution has a width greater than zero. The wider the distribution, the larger the rms voltage level. Hence, unlike the mean voltage, the rms voltage is a useful indicator of the noise level. The rms voltage is of particular usefulness in practical situations because the amount of power associated with a given voltage varies in proportion with the voltage squared. Hence the average <u>power</u> level of some noise fluctuations can be expected to be proportional to v_{rms}^2.

Since thermal noise comes from thermal motions of the electrons we can only get rid of it by cooling the resistor down to absolute zero. More generally, we can expect the thermal noise level to vary in proportion with the temperature.

3.3 'Shot noise'

Many forms of random process produce Gaussian/Normal noise. Johnson noise occurs in <u>all</u> systems which aren't at absolute zero, hence it can't be avoided in normal electronics. Another form of noise which is, in practice, unavoidable is *Shot Noise*. As with thermal noise, this arises because of the quantisation of electrical charge. Imagine a current flowing along a wire. In reality the current is actually composed of a stream of carriers, the charge on each being q, the electronic charge ($1\cdot6 \times 10^{-19}$ Coulombs). To define the current we can imagine a surface through which the wire passes and count the number of charges, n, which cross the surface in a time, t. The current, i, observed during each interval will then simply be given by

$$i = \frac{qn}{t} \qquad \ldots (3.4)$$

Now the moving charges will not be aligned in a precise pattern, crossing the surface at regular intervals. Instead, each carrier will have its own random velocity and separation from its neighbours. When we repeatedly count the number of carriers passing in a series of m successive time intervals of equal duration, t, we find that the counts will fluctuate randomly from one interval to the next. Using these counts we can say that the typical (average) number of charges seen passing during each time t is

$$\langle n \rangle = \sum_{j=1}^{m} \frac{n_j}{m} \qquad \ldots (3.5)$$

19

where n_j is the number observed during the jth interval. The mean current flow observed during the whole time, mt, will therefore be

$$I = \frac{\langle n \rangle q}{t} \qquad \text{...(3.6)}$$

During any <u>specific</u> time interval the observed current will be

$$i_j = \frac{n_j q}{t} \qquad \text{...(3.7)}$$

which will generally differ from I by an unpredictable amount. The effect of these variations is therefore to make it appear that there is a randomly fluctuating noise current superimposed on the nominally steady current, I. The size of the current fluctuation, Δi_j, during each time period can be defined in terms of the variation in the numbers of charges passing in the period, Δn_j, i.e. we can say that

$$\Delta i_j = \frac{q \Delta n_j}{t} \quad \text{where} \quad \Delta n_j = n_j - \langle n \rangle \qquad \text{...(3.8)}$$

As with Johnson noise, we can make a large number of counts and determine the magnitude of the noise by making a statistical analysis of the results. Once again we find that the resulting values have a *Normal* distribution. By definition we can expect that $\langle \Delta n \rangle = 0$ (since $\langle n \rangle$ is arranged to be the value which makes this true). Hence, as with Johnson noise, we should use the mean-squared variation, not the mean variation, as a measure of the amount of noise. In this case, taking many counts and performing a statistical analysis, we find that

$$\langle \Delta n^2 \rangle \approx \langle n \rangle \qquad \text{...(3.9)}$$

Note that — as with the statement that thermal noise and shot noise exhibit Gaussian probability density distributions — this result is based on experiment. In this book we will not take any interest in <u>why</u> these results are correct. It is enough for our purposes to take it as an experimentally verified fact that these statements are true. Combining the above expressions we can link the magnitude of the current fluctuations to the mean current level and say that

$$\langle \Delta i^2 \rangle = \frac{q^2 \langle \Delta n^2 \rangle}{t^2} = \frac{q^2 \langle n \rangle}{t^2} = \frac{q^2}{t^2} \times \frac{It}{q} = \frac{qI}{t} \qquad \text{...(3.10)}$$

Hence we find that the rms size of the random current fluctuations is approximately proportional to the average current. Since some current and voltage is always necessary to carry a signal this noise is unavoidable (unless there's no signal) although we can reduce its level by reducing the magnitude of the signal current.

3.4 An alternative way to describe noise

Up to now we've looked at the statistical properties of noise in terms of its overall rms level and probability density function. This isn't the only way to quantify noise. Figure 3.3 shows an alternative which is often more convenient.

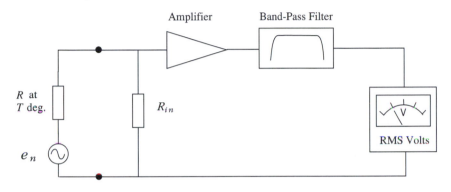

Figure 3.3   Spectral noise measurement.

As in figure 3.1 we're looking at the Johnson noise produced by a resistor. In this case the voltage fluctuations are amplified and passed through a *band-pass filter* to an rms voltmeter. The filter only allows through frequencies in some range, $f_{min} < f < f_{max}$. The filter is said to pass a *bandwidth*, $B = f_{max} - f_{min}$. R_{in} is the input resistance of the amplifier. Note that this diagram uses a common conventional 'trick' of pretending that the noise generated in the resistor is actually coming from an invisible random voltage generator, e_n, connected in series with an 'ideal' (i.e. noise-free) resistor. If we build a system like this we find that the rms fluctuations seen by the meter imply that the (imaginary) noise generator produces an average voltage-squared

$$\langle e_n^2 \rangle = 4kTBR \qquad \qquad \ldots (3.11)$$

where: k is *Boltzmann's Constant* ($=1.38 \times 10^{-23}$ Ws/K); T is the resistor's temperature in Kelvin; R is it's resistance in Ohms; and B is the bandwidth (in Hz) over which the noise voltage is observed. (Note that, as with the earlier statements about *Normal Distribution*, etc, this result is <u>not</u> being proved, but given as a matter of experimental fact.) In practice, the amplifier and all the other items in the circuit will also generate some noise. For now, however, we will assume that the amount of noise produced by R is large enough to swamp any other sources of random fluctuations. Applying Ohm's law to figure 3.3 we can say that the current entering the amplifier (i.e. flowing through R_{in}) must be

$$i = \frac{e_n}{(R + R_{in})} \qquad \qquad \text{... (3.12)}$$

The corresponding voltage seen at the amp's input (across R_{in}) will be

$$v = iR_{in} = \frac{e_n R_{in}}{(R + R_{in})} \qquad \qquad \text{... (3.13)}$$

hence the mean noise <u>power</u> entering the amplifier will be

$$N = \langle iv \rangle = \frac{\langle e_n^2 \rangle R_{in}}{(R + R_{in})^2} \qquad \qquad \text{... (3.14)}$$

For a given resistor, R, we can maximise this by arranging that $R_{in} = R$ when we obtain the *Maximum Available Noise Power*,

$$N_{max} = \frac{\langle e_n^2 \rangle}{4R} \quad \text{which, from eqn 3.8} \quad = kTB \qquad \text{... (3.15)}$$

This represents the highest thermal noise power we can get to enter the amplifier's input terminals from the resistor. To achieve this we have to *match* (i.e. equalise) the source and amplifier input resistances. From this result we can see that the maximum available noise power does not depend upon the value of the resistor whose noise output we are examining.

The *Noise Power Spectral Density* (NPSD) at any frequency is defined as the noise power in a 1 Hz bandwidth at that frequency. Putting $B = 1$ into eqn 3.15 we can see that Johnson noise has a maximum available NPSD of just kT — i.e. it only depends upon the absolute temperature and the value of Boltzmann's constant. This means that Johnson noise has an NPSD which doesn't depend upon the fluctuation frequency. The same result is true of shot noise and many other forms of noise. Noise which has this character is said to be *White* since we the see the same power level in a fixed bandwidth at every frequency.

Strictly speaking, no power spectrum can be truly white over an infinite frequency range. This is because the total power, integrated over the whole frequency range, would be infinite! (Except, of course, for the trivial example of a NPSD of zero.) In any real situation, the noise generating processes will be subject to some inherent mechanism which produces a finite noise bandwidth. In practice, most systems we devise to observe noise fluctuations will only be able to respond to a range of frequencies which is much smaller than the actual bandwidth of the noise being generated. This in itself will limit any measured value for the total noise power. Hence for most purposes we can consider thermal and shot noise as 'white' over any frequency range of interest. However the NPSD <u>does</u> fall away at extremely high frequencies, and this ensures that the total noise power is always finite.

It is also worth noting that electronic noise levels are often quoted in units of *Volts per root Hertz* or *Amps per root Hertz*. In practice, because noise levels are — or should be! — low, the actual units may be $nV/\sqrt{Hz}$ or $pA/\sqrt{Hz}$. These figures are sometimes referred to as the NPSD. This is because most measurement instruments are normally calibrated in terms of a voltage or current. For white noise we can expect the total noise level to be proportional to the measurement bandwidth. The 'odd' units of NPSD's quoted per <u>root</u> Hertz serve as a reminder that — since power $\propto$ volts^2 (or current^2) — a noise level specified as an rms voltage or current will increase with the <u>square root</u> of the measurement bandwidth.

3.5 Other sorts of noise

A wide variety of physical processes produce noise. Some of these are similar to Johnson and shot noise in producing a flat noise spectrum. In other cases the noise level produced can be strongly frequency dependent. Here we will only briefly consider the most common form of frequency-dependent noise: $1/f$ noise. Unlike Johnson or shot noise which depend upon simple physical parameters (the temperature and current level respectively) $1/f$ noise is strongly dependent upon the details of the particular system. In fact the term '$1/f$ noise' covers a number of noise generating processes, some of which are poorly understood. For this form of noise the NPSD, S_n, varies with frequency approximately as

$$S_n \approx f^{-n} \qquad \qquad ...(3.16)$$

where the value of the *index, n,* is typically around 1 but varies from case to case over the range, $0.5 < n < 2$.

As well as being widespread in electronic devices, random variations with a $1/f$ spectrum appear in processes as diverse as the traffic flow in and out of Tokyo and the radio emissions from distant galaxies! In recent years the subject of $1/f$ noise has taken on a new interest as it appears that some *'Chaotic'* systems may produce this form of unpredictable fluctuations.

Summary

This chapter has shown how random noise arises from the quantised behaviour of the real world. Two types of noise — *Johnson Noise and Shot Noise* — were described in detail and their nature shows that they are, in practice, essentially unavoidable. You should now know that noise can only be predicted or quantified on a <u>statistical</u> basis because its precise voltage/current at any future instant is unpredictable. That its magnitude is quantified in terms of averaged rms voltages/currents or mean power levels. The concepts of the *Maximum Available Noise Power* and *Noise Power Spectral Density* were introduced and we saw that Johnson Noise (and also Shot Noise) have a uniform NPSD — i.e. they have a *White* power spectrum. Other forms of noise can show different noise spectra, most commonly a '$1/f$' pattern.

Questions

1) Explain with the help of a diagram how *Thermal Noise* arises. Explain why the mean noise voltage, when averaged over a long time, is almost zero.

2) Explain what's meant by the *Power Spectral Density* of a signal. Thermal and Shot Noise are often said to have a 'white' *Noise Power Spectral Density* (NPSD). What does this tell us about them?

3) A 10 kΩ resistor at 300 K is connected to the input of an amplifier whose input resistance is 22 kΩ. Given that Boltzmann's constant, $k = 1\cdot38 \times 10^{-23}$ Ws/K, calculate the noise power spectral density of the thermal noise the resistor puts into the amplifier. [$3\cdot5\times10^{-21}$ W/Hz.]

4) What value of amplifier input resistance would draw the *Maximum Available Noise Power* from a 10 kΩ resistor? What is the thermal NPSD entering an amplifier with this input resistance when the the 10 kΩ resistor is at 300 K? [10kΩ. $4\cdot12\times10^{-21}$ W/Hz.]

5) How does *1/f noise* differ from *Shot* and *Thermal* noise?

24

Chapter 4

Uncertain measurements

4.1 Doubtful information and errors

Random noise has the effect of making the result of any quantitative measurement uncertain to some extent. This lack of perfect precision is often referred to in terms of producing a given level of *Error* in any result. Alas, many students are rather unhappy with the whole subject of errors. After all, who likes to admit they may have made a 'mistake'? In the minds of many, 'more errors = less marks'! For this reason it's useful to realise that the errors produced by unavoidable random noise <u>aren't</u> something to be embarrassed about. They're a consequence of the real world we're all stuck with. We'll be looking at ways to cope with the effects of noise later on. (We will also see that there are situations where random errors are actually useful!) In this chapter we'll examine how noise affects our ability to communicate information.

To see how noise affects information transmission, consider the situation illustrated in figure 4.1.

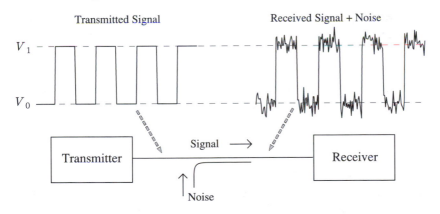

Figure 4.1   Digital communication over a noisy channel.

Here a message is being sent as a stream of binary digits, i.e. it is in the form of a *Serial Digital* signal. The transmitter uses one voltage level, V_1, to signal a '1' and another voltage, V_0, to signal a '0'. The information is therefore carried by the voltage pattern. Some random noise is introduced during transmission. As a result, the received signal is a combination of the intended signal voltage pattern and this added noise. For simplicity we can assume that the transmitter and receiver are in themselves 'perfect', i.e. they don't generate any noise of their own. In reality this won't be true. For our purposes here it doesn't really matter where the noise comes from. Any actual noise coming from the transmitter/receiver circuits would have an identical effect to the same total noise level injected onto the channel from an external source.

In the absence of any noise the receiver could repeatedly measure the input it sees and decide, "If this is V_0 I've received a '0', if it's V_1 I've received a '1'." However, the noise means that the input it sees is hardly ever actually equal to V_0 or V_1. It therefore requires some other recipe for deciding whether it's received a '0' or '1'. The simplest way to do this is to define a sensible *Decision Level, V'*, mid-way between V_0 and V_1

$$V' \equiv \frac{V_0 + V_1}{2} \qquad \dots (4.1)$$

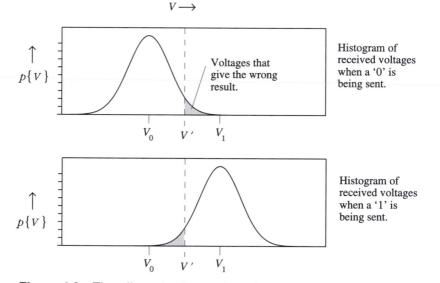

Figure 4.2 The effect of noise on the voltages seen by the receiver.

The receiver now works by saying, 'If I see a voltage $\geqslant V'$ I've received a '1', if I see a voltage $< V'$ I've received a '0'.'

The results of decoding a noisy digital message in this way can be understood by looking at figure 4.2.

The effects of the noise can be assessed by making a large number of measurements of the received voltage levels and plotting a probability distribution of the results. The top graph shows a plot of the distribution of voltages seen by the receiver when the transmitter is sending V_0. In this situation the received voltage will be $V_0 + v_n$ where v_n varies randomly from one measurement to another. Since the <u>average</u> noise voltage of lots of measurements is essentially zero the resulting spread of voltages has its mean at V_0. For *Normal* noise the distribution therefore has a Gaussian shape with its peak at V_0. A similar result, shown in the lower graph, arises when the transmitter is trying to send V_1, but in this case the average (and peak of the shape) are at V_1.

Since the receiver decides that any voltage above V' is a '1' and any voltage below V' is a '0' we can predict the frequency of mistakes by calculating the fraction of the plots which are the wrong side of V'. When the transmitter is trying to send V_1 the probability or relative frequency, $p\{V\}$, with which the received voltage is seen to be in a small interval, dV, centred at some voltage, V, will be

$$p\{V\}.dV = A. \operatorname{Exp}\left\{\frac{-2(V - V_1)^2}{\sigma^2}\right\}.dV \qquad \text{... (4.2)}$$

Since the observed voltage <u>must</u> always be somewhere in the range from $-\infty$ to $+\infty$ we can say that the value of the coefficient, A, must be such that

$$\int_{-\infty}^{+\infty} A. \operatorname{Exp}\left\{\frac{-2(V - V_1)^2}{\sigma^2}\right\} dV = 1 \qquad \text{... (4.3)}$$

i.e. the probability that the observed voltage is somewhere between $-\infty$ and $+\infty$ is unity. Since the total area under the distribution shape isn't affected by the choice of V_1, this is equivalent to saying that

$$\frac{1}{A} \equiv \int_{-\infty}^{+\infty} \operatorname{Exp}\left\{\frac{-2V^2}{\sigma^2}\right\} dV \qquad \text{... (4.4)}$$

When V_1 is being sent, the chance, C_1, it will be correctly received is determined by the fraction of the distribution which lays above V'. This can be determined from integrating over the appropriate part of the curve to obtain

$$C_1 = \int_{V'}^{\infty} p\{V\} \, dV = \int_{V'}^{\infty} A. \operatorname{Exp}\left\{\frac{-2(V - V_1)^2}{\sigma^2}\right\} dV \qquad \text{... (4.5)}$$

In a similar way, the chance C_0, that V_0 will be correctly received is determined by the fraction of the distribution which is below V' when V_0

is being sent, i.e. we can say

$$C_0 = \int_{-\infty}^{V'} p\{V\}\, dV = \int_{-\infty}^{V'} A.\,\mathrm{Exp}\left\{\frac{-2\,(V_0 - V)^2}{\sigma^2}\right\} dV \qquad \text{... (4.6)}$$

Using a book of standard integrals we can find that the above expressions are equivalent to

$$C_1 = \frac{1}{2}\cdot\left[1 + \mathrm{Erf}\left\{\frac{\sqrt{2}.\,(V_1 - V')}{\sigma}\right\}\right] \qquad \text{... (4.7)}$$

$$C_0 = \frac{1}{2}\cdot\left[1 + \mathrm{Erf}\left\{\frac{\sqrt{2}.\,(V' - V_0)}{\sigma}\right\}\right] \qquad \text{... (4.8)}$$

and

$$A \equiv \left(\frac{1}{\sigma}\right).\sqrt{\frac{2}{\pi}} \qquad \text{... (4.9)}$$

where Erf is a standard mathematical function called the *Error Function*. Since this isn't a pure maths book the details of this proof and the precise nature of the error function don't matter very much. It is enough for us to accept that it is just another function like sine or cos that we can look up in a book and which happens to be the right one to solve the integrals. We can now use the above expressions to see how often the receiver will pick up the correct signal level in the presence of some noise.

Since we defined V' to be mid-way between V_0 and V_1 we have a situation where $C_1 = C_0$. Hence we only need to look at how one of the above depends upon the chosen voltages and the noise level. The amplitude of the signal voltage being transmitted is $V_s = V_1 - V_0$. The rms amplitude of the typical noise voltage is σ. Since $V' = (V_1 + V_0)/2$ we can therefore say that the fractional chance of each '1' or '0' being received correctly will be

$$C = \frac{1}{2}\cdot\left[1 + \mathrm{Erf}\left\{\frac{V_s}{\sqrt{2}.\,\sigma}\right\}\right] \qquad \text{... (4.10)}$$

The dependence of C upon the signal/noise voltage ratio, V_s/σ, can be seen by looking at the curve plotted in figure 4.3. As we would expect C approaches unity when the signal to noise ratio is high. In this situation the signal voltage is very big compared to the noise, hence the noise will have no noticeable effect.

A more curious result is that when the signal/noise ratio is zero $C = 0.5$ — i.e. the receiver will correctly pick up 50% of the message's pattern of '1's and '0's even when the transmitter doesn't send the signal! At first sight this seems very odd. Surely, if the signal amplitude is zero the receiver has no way to know what the message is...

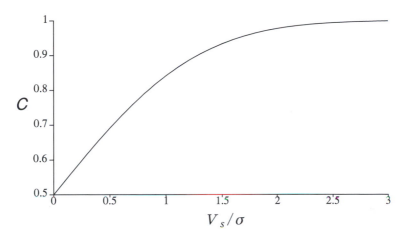

Figure 4.3 Fraction of '1's and '0's received correctly.

The reason for this odd result can be explained as follows. Imagine that we didn't bother with using a proper signal receiver but instead just kept tossing a coin. Every time we get a head we decide the message should contain a '1'. Every tail is taken as a '0'. In this way we can build up a pattern of '1's and '0's without bothering to look at the actual signal. Since there are only two possibilities ('1' or '0'), every time we throw the coin we have a 50% chance of getting the correct result. As a consequence 50% of the '1's and '0's in our coin-generated version of the message will be correct. However, this <u>doesn't</u> mean that we have received 50% of the actual information since we don't know <u>which</u> 50% of the coin-generated bits are the correct ones! This result is just the same as if we'd used random noise to make the receiver perform the equivalent of 'toss coins' to generate a random string of bits.

This demonstrates an important feature of the way information is communicated and processed. The amount of information we have doesn't just depend upon how many bits we've gathered. It also depends upon how confident we are that each bit is correct. The amount of information received depends upon how <u>certain</u> we are that the pattern is correct. If we're only 50% certain and there are only 2 possibilities we don't actually have <u>any</u> real information since any other outcome is just as likely to be the correct one.

29

Summary

This chapter has shown how the effect of noise is to produce random errors when we communicate a signal. These random errors mean we can never be absolutely certain that we've received the correct information. Since noise is present in all real systems this means that we can never be certain that the information we have is absolutely correct. You should also now know that the amount of information in a signal pattern depends upon how certain we are that it is correct.

Questions

N.B. In the following, use the approximation

$$\text{Erf}\{x\} \approx 1 - \frac{0 \cdot 348t - 0 \cdot 0958t^2 + 0 \cdot 748t^3}{\text{Exp}\{x^2\}}$$

where

$$t \equiv \frac{1}{1 + 0 \cdot 47x}$$

1) A digital transmission system uses $0 \cdot 5$ V to signal a logical '0' and $4 \cdot 5$ V to signal a logical '1'. A message is transmitted which consists of a sequence which contains 2000 '1's and 2000 '0's. The channel used to carry the message has a noise level we can characterise by a value of $\sigma = 1 \cdot 5$ V. How many bits are likely to be received correctly using a receiver whose decision level is set mid-way beween the logical '0' and '1' levels? **[3984 in total.]**

2) How many bits would have been received correctly in question 1 if the receiver's decision level had been set at either *a)* 3 V, or *b)* 1 V? (Remember that a chance of a '0' being received correctly is C_0 and the chance of a '1' being received correctly is C_1.) **[a) 3953, b) 3494.]**

3) Write a program to calculate how the chance of correct reception in the system described above varies if the decision level is varied between $0 \cdot 5$ V and $4 \cdot 5$ V.

Chapter 5

Surprises and redundancy

In the last chapter we saw how the amount of information in a received signal pattern depends upon how confident we can be that its details are correct. We also saw how the probability that a digital bit of information will be correctly received is

$$C = \frac{1}{2}\left[1 + \text{Erf}\left\{\frac{V_s}{\sqrt{2}.\sigma}\right\}\right] \qquad \ldots (5.1)$$

where V_s is the peak-to-peak size of the signal voltage and σ is a measure of the width of the noise voltage's probability density pattern (histogram). This expression is theoretically fine, but it can be awkward to use in practice. In most real situations it is more convenient to deal with signal and noise <u>powers</u> or rms voltages. We therefore need to turn expression 5.1 into a more useful form.

A square-wave of peak-to-peak amplitude, V_s, will have a mean power $S = (V_s/2R)^2$ where R is the appropriate resistance across which the observed voltage appears. Hence we can use

$$V_s = 2\sqrt{RS} \qquad \ldots (5.2)$$

to replace the signal voltage in the above expression. To establish the noise power in terms of the width, σ, we have to evaluate the noise's rms voltage level, v_{rms}. To do this we can argue as follows:

In chapter 3 we saw that the noise level can be represented in terms of a probability distribution of the form

$$p\{V\}.dV = A\,\text{Exp}\left\{-\frac{2V^2}{\sigma^2}\right\}.dV \qquad \ldots (5.3)$$

where, to ensure that the actual voltage always lies between $+\infty$ and $-\infty$, we can say that

$$A \equiv \frac{1}{\sigma}\sqrt{\frac{2}{\pi}} \qquad \ldots (5.4)$$

To compute the rms voltage we take many voltage readings, square them, add them together, divide by the number of readings, and take the square root of the result. This is mathematically equivalent to

$$v_{rms} = \sqrt{\int_{-\infty}^{+\infty} V^2 p\{V\}\, dV} \qquad \qquad \dots (5.5)$$

This is because, when we make lots of voltage measurements, the fraction of them which falls between V and $V + dV$ is $p\{V\}.dV$. By solving the above integral we discover that

$$v_{rms} = \frac{\sigma}{2} \qquad \qquad \dots (5.6)$$

The noise power level will be $N = v_{rms}^2 / R$ and hence we can say that

$$\sigma = 2\sqrt{RN} \qquad \qquad \dots (5.7)$$

We can now use 5.2 and 5.7 to replace V_s and σ in expression 5.1 and obtain the result

$$C = \frac{1}{2}\left[1 + \text{Erf}\left\{\sqrt{\frac{S}{2N}}\right\}\right] \qquad \qquad \dots (5.8)$$

This expression tells us the chance that bits will be received correctly in terms of two easily measurable quantities — the signal power, S, and the noise power, N.

Whenever possible we should make the signal/noise power ratio as large as we can to minimise the possibility of errors. If this is done we can often neglect the information loss produced by random noise. It should be remembered, however, that the signal/noise ratio will always be finite. Hence we can never get rid of this problem altogether. Despite this, a *S/N* power ratio of just 10 gives $C = 0.999214$ — i.e. around 99·98% of the bits in a typical message received with this *S/N* would be correct. A slightly better *S/N* ratio of 25 gives $C = 0.9999997$. This is equivalent to an *Error Rate* of around 3 bits in every ten million (3:10,000,000).

It may seem that an error rate below 'one in a million' isn't really worth making a fuss about. Alas, there are some factors which we have not, as yet, taken into account.

> • We will usually be sending a number of bits to indicate a code *word* and these words may be built up into a longer message. A given message may be composed of a lot of bits.
> • Whilst a 1:1,000,000 error rate may be acceptable for many purposes, it may be a disaster in other circumstances.

For example, consider one of the systems used to signal to strategic defence nuclear submarines. These submarines are designed to cruise hidden below the ocean surface. They should remain hidden up until such time as they might be required to launch a nuclear attack. For this reason they avoid transmitting any radio signals which would give away their location to an enemy. Their standing orders include an instruction to 'launch retaliation' if their home county is destroyed by a 'sneak attack'. The question therefore arises, 'How can they tell if their home country has been flattened?'. A country that has been destroyed may not have any radio transmitters left to transmit a signal to the subs, ordering them to attack.

To get around this problem the military devised a 'fail disaster' system. The home country regularly transmits a sequence of coded messages at prearranged moments. The sub pops up a radio buoy at these times, listens for these broadcasts, and verifies that the codes are correct. The submarine commander then uses the <u>absence</u> of these messages as a 'signal' to the effect that, 'Home has been wiped out, attack enemy number 1'. An incorrectly coded message is interpreted as a 'signal' that, 'We have been taken over by an enemy and forced to make this broadcast against our will. Attack!'. Clearly, for a signalling system of this kind a single error could be a genuine disaster. Even a one in a million chance of a mistake is far too high. So steps have to be taken to make an error practically impossible. The importance of errors varies from situation to situation, but it should be clear from the above example that we sometimes need to ensure very low error rates.

When dealing with the effects of errors on <u>messages</u> (rather than on single bits) we must also take into account how effectively we are using our encoding system to send useful information and how important the messages are.

Some messages are quite surprising, whereas others are so predictable that they tell us almost nothing. To quote some examples from the English language.

    1) 'This car does 0 – 60 in 0·6 seconds.'
    2) 'If you want to catch a bus you should q over there.'
    3) 'Party at 8, bring a bottle.'

The contents of this first message indicate a remarkable car! Although every symbol in this sentence looks OK by itself, the whole message is clearly rather suspect. We can only guess what the correct message was. In the second message the error is pretty obvious and we can feel almost certain that we know what the correct message should be. The third message looks fine, but it may still be wrong, e.g. the party may be at 9 o'clock and the figure 8 is, in fact, a mistake. It is also ambiguous; the party may be at house <u>number</u> 8, not at 8 o'clock.

Messages 1 and 3 contain examples of errors which, if noticed, give no reliable indication of the correct message. In order to deal with errors of this type we need to include some <u>extra</u> information in the message.

Message 2 contains an error which can be corrected. In the context of the message and our knowledge of the English language, the whole word 'queue' is unnecessary. The 'ueue' is *redundant*. This relates to the observation that — in English — the letter 'u' is often redundant. We could replace almost every 'qu' in English with 'q' without the correct meaning being lost (although we'd get complaints about our spelling!)

Clearly, it is valuable to choose a system of coding which makes errors obvious and allow us to correct received messages. To see how it is possible to produce systems which do this we need to analyse redundancy and its effect on the probability of a message being understood correctly.

In an earlier chapter we saw that the amount of information in a message can be expected to increase with $\log_2$ of the number of code symbols available. This, in fact, assumes that all the available symbols are used (a symbol which isn't used might as well not exist). It also assumes they are all used with similar frequency. Hence the probability, P, of a particular symbol appearing would $= 1 / M$ where M is the number of available code symbols. We can therefore say that the amount of information would vary with

$$\log_2 \{M\} \;=\; \log_2 \left\{ \frac{1}{P} \right\} \;=\; - \log_2 \{P\} \qquad \ldots (5.9)$$

Consider the situation where we use a set of M symbols, $X_1, X_2, X_3, \ldots X_M$, for sending messages. By collecting a large number of messages and examining them we can discover how often each symbol tends to occur in a typical message. We can then define a set of probability values

$$P_i \equiv \frac{N_i}{N} \qquad \ldots (5.10)$$

from knowing that each symbol, X_i , occurs N_i times in a typical message N symbols long. In a situation where all the symbols tend to appear equally often we can expect that $P_i = \frac{1}{M}$ for every symbol X_i — i.e. all the symbols are equally probable. More generally, the symbols appear with various frequencies and each P_i value indicates how often each symbol appears.

When all the symbols are equally probable the amount of information provided by each individual symbol occurrence in the message will be $\log_2\{M\}$. The <u>total</u> amount of information in a typical message N symbols long would then be

$$H = N.\log_2\{M\} = -N.\log_2\{P\} \qquad \text{...}(5.11)$$

where $P = P_i$ for all i, since in this case the probabilities all have the same value. This expression giving the total amount of information in terms of symbol probabilities indicates how we can define the amounts of information involved when the symbols occur with differing frequencies. We can then say that the amount of information provided just by the occurrences of, say, the X_i symbol will be

$$H_i = -N_i\log_2\{P_i\} = -NP_i\log_2\{P_i\} \qquad \text{...}(5.12)$$

From this expression we can see that the smaller the probability of a particular symbol, the more informative it will be when it appears. Surprising (i.e. rare) messages convey more information than boringly predictable ones! The total amount of information in the message will therefore be the sum of the amounts carried by all the symbols

$$H = \sum_{i=1}^{M} -N_i\log_2\{P_i\} = \sum_{i=1}^{M} -NP_i\log_2\{P_i\} \qquad \text{...}(5.13)$$

From expression 5.12 we can say that every time X_i appears in a typical message it provides a typical amount of information <u>per symbol occurrence</u> of

$$h_i = H_i/N_i = -\log_2\{P_i\} \qquad \text{...}(5.14)$$

(Note that to make things clearer we will use H to denote total amounts of information and h to denote an amount per individual symbol occurrence.) From expression 5.14 we can say that the amount of information per symbol occurrence, averaged over <u>all</u> the possible symbols is

$$h = \frac{H}{N} = \sum_{i=1}^{M} -P_i\log_2\{P_i\} \qquad \text{...}(5.15)$$

In general, this averaged value will differ from the individual h_i values <u>unless</u> all the symbols are equally probable. Then $P_i = 1/M$ and 5.15 would become equivalent to $h = M\times(-(1/M)\log_2\{P_i\}) = h_i$. It's interesting to note that the form of the above expressions is similar to those used for entropy in thermodynamics. Many books therefore use the term *entropy* for the measure of information in a typical message or code.

The above argument gives us a statistical method for calculating the amount of information conveyed in a typical message. Of course, some messages <u>aren't</u> typical, they're surprising. The information content of a <u>specific</u> message may be rather more (or less) than is usual. The above expressions only tell us the amounts of information we tend to get in an average message.

Consider now a specific message N symbols long where each symbol, X_i, actually occurs A_i times. The amount of information provided by each individual symbol in the message is still $-\log_2\{P_i\}$, but there are now A_i of these, not the NP_i we would expect in a 'typical' or average message. We can therefore substitute A_i into expression 5.13 and say that the total amount of information in this <u>particular</u> message is

$$H = -\sum_{i=1}^{M} A_i . \log_2\{P_i\} \qquad ...(5.16)$$

In order to convey information, every one of the symbols we wish to use must have a defined meaning (otherwise the receiver can't make sense of them). This is another way of saying that the number of available symbols, M, must always be finite. Since any particular symbol in a message must be chosen from those available we can say that

$$\sum_{i=1}^{M} P_i = 1 \qquad ...(5.17)$$

In most cases the chance of a particular symbol occurring will depend to some extent upon the previous symbol (e.g. in English, a 'u' is much more likely to follow a 'q' than any other letter) and some combinations of symbols occur more often than others (e.g. 'th' or 'sh' are more common than 'xz'). The term *Intersymbol Influence* is used to describe the effect where the presence (or absence) of some symbols in some places affects the chance of other symbols appearing elsewhere. To represent this effect we can define a *Conditional Probability*, $P_{i \to j}$ to be the probability that the j th symbol will follow once the i th has appeared. The chance that the symbol combination X_iX_j will appear can then be assigned the *Joint Probability*,

$$P_{ij} = P_i . P_{i \to j} \qquad ...(5.18)$$

Just as the amount of information provided by an individual symbol taken by itself depends upon its probability, so the extra information provided by a following symbol depends upon how likely it is once the previous symbol has already arrived. For example — in English, a 'u' is a virtual certainty after a 'q', hence it doesn't provide very much extra information once the 'q' has arrived. The 'u' is said to be redundant once the 'q' has arrived. However, although it doesn't provide any real extra information it <u>is</u> useful as a way of checking the correctness of the received message.

The term *Conditional Entropy* is often used to refer to $h_{i \to j}$, the average amount of information which is communicated by the jth symbol after the ith has already been received. Since h is proportional to $\log_2 \{P\}$, the *joint entropy*, h_{ij} (the amount of information provided by this pair of symbols taken together), must simply be

$$h_{ij} = h_i + h_{i \to j} \qquad \qquad \text{...(5.19)}$$

If we wish to maximise the amount of information in a typical message then we would like every symbol and combination of symbols to be as improbable as possible (i.e. minimise all the P values). Alas, expression 5.17 means that when we make one symbol or combination <u>less</u> likely some others must become <u>more</u> probable. We can't make all the existing symbols less likely without adding some new ones! From the English language example of a 'u' following a 'q' we can see that the effect of intersymbol influence is generally to <u>reduce</u> the amount of information per symbol since the 'u' becomes pretty likely after a 'q'. Hence we can expect that the information content of a message is maximised when the intersymbol influence is zero. Under these conditions

$$h_{ij} = h_i + h_j \quad \text{(no influence)} \qquad \qquad \text{...(5.20)}$$

i.e. the amount of information communicated by two symbols is simply double that provided by either of them taken by itself. In such a situation none of the transmitted symbols are redundant. Since this is the best we can do, it follows that, more generally

$$h_{i \to j} \leqslant h_j \qquad \qquad \text{...(5.21)}$$

i.e. the average extra information produced by the following symbol can never exceed that which it would have as an individual if there were no intersymbol influence.

Summary

You should now understand how the amount of information in a message depends upon the probabilities (or typical frequencies of occurrence) of the various available symbols. That the chance of transmission errors depends upon the signal/noise ratio. That the amount of information in a <u>specific</u> message can differ from an average one depending upon how surprising it is (how many times specific symbols actually occur in it compared with their usual probability). That *Intersymbol Influence* can help us check that a message is correct, but reduces the maximum information content.

Questions

1) An information transmission system uses just 4 symbols. The symbols appear equally often in typical messages. How many bits of information does each symbol carry? How much information (in bits) would a typical message 1024 symbols long contain? [**2 bits per symbol. 2048 bits.**]

2) An information transmission system uses 6 symbols. Four of these, $X_1 X_2 X_3$ and X_4, have a typical probability of appearance, $P = 0.125$. The other two symbols, X_5 and X_6 have probabilities, $P = 0.25$. How much information would a typical message 512 symbols long carry? How much information would a <u>specific</u> message 512 symbols long carry if it only contained 300 X_1's, 100 X_3's, and 112 X_6's? [**1280 bits. 1424 bits.**]

3) Explain what's meant by the term *Intersymbol Influence*. Say why and when this can be either a 'good thing' or a 'bad thing' depending upon the circumstances.

Chapter 6

Detecting and correcting mistakes

6.1 Errors and the law!

In chapter 4 we saw that random noise will tend to reduce the amount of information transmitted or collected by making us uncertain that the resulting message pattern is correct. We've also seen how *redundancy* can provide a way to check for mistakes and, in some cases, correct them. One of the advantages of digital signal processing systems is that they are relatively (but <u>not</u> totally) *immune* from the effects of noise. A S/N ratio of just 10:1 is enough to ensure that 99·92% of digital bits will be correct.

For short, unimportant, messages this level of immunity from errors is fine, but it isn't good enough for other situations. For example, consider a computer which has to load (read) a 200 kbyte (1·6 <u>million</u> bits) wordprocessing program from a disc. A 0·01% *error rate* would mean the loaded program would contain around 160 mistakes! This would almost certainly cause the program to crash the computer. By the way, note that the term 'error rate' <u>doesn't</u> mean the errors appear at regular intervals. If it did, we could simply count our way along the pattern to find and correct the errors! The errors will be randomly placed. The rate simply indicates what fraction of the bits are likely to be wrong, not where they are. The term, 'error rate' is therefore potentially misleading, although it is commonly used.

We can reduce the rate at which errors occur by improving the S/N ratio, but there is, in fact, a better way, based on deliberate use of redundancy. By introducing some *intersymbol influence* we can make some patterns of symbols *illegal* — i.e. we arrange that they can <u>only</u> occur as the result of a mistake. This makes it possible to detect that the signal pattern contains an error. The main disadvantage of this technique is that we have to reduce the amount of information we're trying to get into a given message since some of the symbols are now being used to 'check' others rather than sending any extra information of their own. (It can be argued that this doesn't really matter since — if we don't do anything about it — random noise will destroy some of the information anyway, although we may not know about it!) One of the simplest ways to deal with errors is to repeat the message. The two versions can then be compared to see if

39

they're the same.

If the probability that any particular bit or symbol in a message is correct is C, then the chance that it's an error must be $E = 1 - C$. (It must be either right or wrong!) As a result, when we send and compare two copies of a message, the chance that <u>both</u> copies have a symbol error in the same place will be E^2. As an example consider a system whose S/N ratio provides a chance $C = 0.999$ that individual bits are correct. This means that $E = 0.001$ per bit. The chance of both copies of a specific bit being wrong will therefore be $E^2 = 0.000001$ — i.e. in a typical pair of repeated messages there is only a 1:1000000 chance that both copies of any particular bit will be wrong. Now the chance that a particular bit in 'copy #1' is correct and 'copy #2' is wrong will be $C.E = 0.000999$. Similarly, the chance that just the first copy is wrong will be $E.C = 0.000999$. The chance that both are correct will be $C^2 = 0.998001$.

When we compare two versions of a long message we therefore typically find that $100 \times (C^2 + E^2)\% = 99.8002\%$ of the bits agree with their copies and just $100 \times 2 \times (C.E)\% = 0.1998\%$ differ. As a result we can see that just under 0.2% of the bit 'pairs' disagree. We have <u>detected</u> the presence of the errors which caused these disagreements and know where in the message they appear. This is the advantage of this 'repeat message' technique over just sending one copy. In this example, sending two copies is 'redundant' because they should both contain the same information. Once we know about the errors we can take appropriate action (e.g. ask the transmitter to repeat the 'uncertain' parts of the message). A single copy of the message would contain about 0.1% mistakes <u>but we wouldn't know about them</u> unless we arrange for some redundancy. Hence without redundancy we can't do anything to recover what we've lost.

The system of using a pair of messages isn't perfect. (What is?) There are still E^2 errors which we <u>won't</u> spot because both copies have been changed in the same place. As a result there are still 0.0001% <u>undetected</u> errors in the received information. However, this is much better than the $E = 0.1\%$ of undetected errors we'd get if only one copy of the message had been sent.

By spotting differences between two copies of the message we can detect nearly all of the places where there has been a random noise produced error. However, we still don't know which of the differing versions is correct. A way to overcome this is to go one stage further and use the military approach called, *Tell Me Three Times*. This means we send <u>three</u>

copies of the same message. Using the same arguments as before we can now say the chance that all three copies of a specific bit are correct is $C^3 = 99.70\%$. The chance that any one version is wrong is $3C^2E = 0{\cdot}2994\%$. (There are three chances for one version out of three to be wrong.) Similarly, the chance that two versions both have an error in any specific bit is $3C.E^2 = 0{\cdot}0002997\%$. The chance that all three are wrong is $E^3 = 0{\cdot}0000001\%$. (N.B. These values have all been rounded to 4 significant figures to make them more readable!)

One effect of tell me three times is to reduce the undetected error rate (E^3) still further. However, the main benefit is that nearly all the errors can now be *corrected*. This is because in most cases a difference between the three versions of the message occurs because just one of them is wrong. The signal receiver can therefore work on a 'majority vote' system and decide that, 'when two versions agree and one differs, the correct signal is the one shown by the two versions in agreement'. It then can use this rule to recover the 'correct' information. Occasionally, this means it will make a mistake when two versions have been changed by errors, but from the figures shown above we can see this will only happen for about one correction in a thousand. Hence the tell me three times technique allows us to detect and correct most of the errors produced by random noise.

6.2 Parity and blocks

The disadvantage of tell me three times is that we have to send every message three times instead of being able to send three different sets of information with the same number of bits or symbols. This makes it a relatively inefficient and slow way to convey (or store) information. Fortunately, there are various other methods available for detecting and correcting errors which don't reduce the overall information carrying capacity quite so much. One of the most common digital techniques is the use of *Parity* bits. Before explaining these it's useful to consider the concept of binary *Words*.

From previous chapters you should already be familiar with the idea of using a set of binary digits (*bits*) to represent information. (See, for example, chapter 1 where we represented a series of sampled voltages as '000', '001', etc.) It's usual to refer to groups of eight bits as a *Byte* of information. This stems from early computers which mostly handled 8 bits of information at a time. More generally, the term *word* has come to mean a group of bits which carry a convenient amount of information. Most modern desktop computers have microprocessors which can handle 16 or

32 bits of information at a time. The information in such a processing system is said to be held as a set of 16 or 32 bit *words*. Each binary word can then be regarded as a digital *Symbol*. These symbols can be built up in patterns to represent the information. Unlike the term 'byte', 'word' can mean any convenient number of associated bits.

To see how *Parity Checking* can be used to detect and correct errors, imagine a system when the information is initially held as a series 8-bit words. The system may want to transmit — or process in some other way — a series of words, %10011100, %10010100, %11100101, etc. (Note that here a '%' before the number is used to indicate that it's in binary notation.) The parity of each word can be defined to be odd or even depending upon how many '1's it contains. On this basis, %10011100 has even parity, %1101010 has odd parity, %11100101 has odd parity, and so on. We can now add an extra bit onto each word to represent its parity. For example, we can add a '1' onto the end if the word was even or a '0' onto the end if it was odd. This converts the initial words as follows:

$$\%10011100 \Rightarrow \%100111001$$
$$\%10010100 \Rightarrow \%100101000$$
$$\%11100101 \Rightarrow \%111001010$$

We now transmit or process these new <u>9-bit</u> words instead of the original 8-bit ones. This extra bit we've tacked onto each original word doesn't carry any fresh information. It's called a *parity bit* because it simply confirms the parity of the other bits in the word. This means the patterns we transmit are now partially redundant and this redundancy can be used by the receiver to check for errors. Under the system we've chosen <u>every</u> *legal* 9-bit word has an odd number of '1's. The receiver can now read each 9-bit word as it arrives and check that it's parity is, as expected, odd.

Random noise may occasionally change one of the bits in a word during transmission. As a result, the received 9-bit word will now have an even number of '1's. The receiver can spot this fact and use it to recognise that the word is *illegal*. This means that it's not a pattern which the transmitter would send. Hence the receiver can discover that it must contain an error. In this way the parity bits allow error detection. Note that this isn't the only way to implement parity bits. We could put the extra bits at the start of the words, or somewhere in the middle. We could choose to add a '1' onto the odd words and a '0' to the even ones to make all the legal 9-bit words have even parity. The details don't matter so long as the receiver knows what to expect.

Using this example we can now define some ways to quantify the degree of redundancy in the coding system used to transmit information. In this case, each 9 transmitted bits only contains 8 bits worth of real information. We can define the ratio of number of bit of information to the number of bits transmitted to be the *Efficiency* of the coding system used, i.e. we can say

$$\text{Efficiency} = \frac{\text{number of information bits}}{\text{number of info bits} + \text{number of parity bits}} \qquad \text{... (6.1)}$$

In this case the ratio is 8/9, hence the transmission system has an efficiency of 0·888. The *redundancy* can be defined to be one minus the efficiency, $1 - \frac{8}{9} = 0\cdot111$. These values can be compared with the 'tell me three times' system where we had to send three times as many bits as were required to contain the original information — i.e. an efficiency of 1/3rd or redundancy of 2/3rds. Note that although the parity system we've described is more efficient than 'tell me three times' it still requires us to send more bits than were needed for the original information. This is a general rule. Every system for detecting (and correcting) mistakes produced by random noise requires us to communicate or store 'extra' bits which essentially repeat some of the information.

Comparing the parity checking system described above with 'tell me three times' we can see it can detect occasional 1-bit errors, but has a much lower redundancy. However, it can't correct errors. To do that we can use a slightly more complex approach based upon what are called *Block Codes*. A simple example is shown below. Here we collect the words we want to transmit into a series of blocks of the kind illustrated, e.g. a data stream %01011000, %11100011, %00011011, %11001100, %010..., etc. is collected into blocks of four words to make patterns of 8×4 bits like:

```
                                'row' parity bits
        %01011000      →        0
        %11100011      →        0
        %00011011      →        1
        %11001100      →        1
'column' parity bits           ↓
        %10010011
```

We now generate a set of 'row' parity bits for checking each words. We also generate a set of 'column' parity bits — using the first bit of each

43

number for the first parity bit, then the second bit of each number for the second, etc. In the example we've chosen this means that each original block of 8×4=32 bits of information is used to produce an extra 12 bits. We then transmit all 44 bits to a receiver. To see what happens when a random error occurs during transmission we can assume that the received version of the above turns out to be as shown below

		Received	Computed
%01011000	→	0	0
%11100*1*11	→	0	*1*
%00011011	→	1	1
%11001100	→	1	1

↓

Received	%10010011
Computed	%10010*1*11

The receiver collects the received block of data and parity bits sent by the transmitter. It then computes its own version of what the parity bits should be and compares them with the values it has received. In this example one of the bits has been altered from a '0' to a '1' during transmission. As a result, the received and computed parity bits won't agree and the receiver can tell that there's a mistake in the block it has received. It can now use one parity disagreement to identify which row the error is in and the other to identify the column it is in. As a result it can locate and correct the mistake. This ability of block codes to both detect and correct mistakes is an important feature of modern information processing.

Note that there's nothing magic about the choice of choosing an 8×4 block size. We could have arranged the block as 8×8, or put two words on each row and used 16×16, or even split the words to make some peculiar arrangement like 11×7. Provided the transmitter and receiver use the same rules any arrangement may be OK. Note also that we aren't limited to a 'two-dimensional' block. We could arrange the bits in a 'cube' of, say, 8×8×8 bits, and collect a third set of parity bits running through the pattern in another 'direction'. (In principle, we can arrange the bits in a many-dimensional pattern although it gets a little hard to visualise!)

The choice of block arrangement depends upon how worried we are about the effects of noise. The 2-dimensional example shown above works fine for single bit errors, but runs into trouble if there is more than one error in a block. For example, if there are two errors in a row then the received and computed parity values for that row will agree. The receiver would then be able to detect that two columns contained errors, but not which row they were on. Hence this simple example can correct 1-bit errors but only detect 2-bit errors in a block.

In general, the error detecting and correcting ability of a block code can be defined in terms of measure called the *Minimum Hamming Distance*. Block codes work because some transmitted word patterns of '1's and '0's are illegal. The *Hamming Distance* between any pair of legal words is defined as the number of bits which have to be changed to convert one word into the other. The *Minimum Hamming Distance* is defined as the lowest Hamming Distance value we find between <u>any</u> pair of legal words in the chosen code system. This provides us with a number which determines how well a code system can cope with errors.

The properties of well designed code systems with various Minimum Hamming Distances are as follows:

MHD = 1	No error immunity (every pattern appears legal)
MHD = 2	Detects 1 error, no correction
MHD = 3	Detects and corrects 1 error
MHD = 4	Detects up to 2 errors and can correct 1 error
MHD = 5	Detects 2, corrects 2
etc...	

In a given situation we can start by deciding how many errors at a time we want to be able to spot or correct. Then use the MHD to tell us how many illegal patterns have to 'surround' each legal one. This then tells us how much redundancy and how many parity bits we need.

6.3 Choosing a code system

There is an enormous variety of data encoding systems. It sometimes seems as if theoreticians keep inventing new ones purely as something to name after themselves! Despite this, many of them are designed to have features useful in specific situations. Most are designed to combat random errors and work along the lines described in the last section. We will be looking at an example of a powerful error correcting code when we examine how Compact Discs work in a later chapter. Here we will examine two special systems which have properties useful for particular jobs. The first example is a digital *Linear Encoder*.

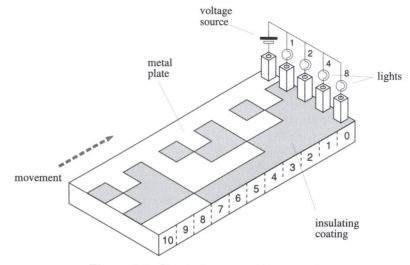

Figure 6.1 Simple linear position encoder.

Figure 6.1 shows a metal plate which has a pattern of insulating material placed upon its surface. A line of electrical contacts is arranged to press against the plate. When they touch the metal a current can flow through them. This current will be blocked if they are touching a part of the surface coated with insulator. The contacts therefore form a set of sensors which produce a pattern of currents which changes in response to plate movements. In the figure these currents are shown connected to a row of lights which would light up to indicate the plate position. More commonly, the sensors would be connected to a computer system to input a binary number which represents the position of the plate. Hence the system acts as an encoder which provides a signal which changes as the plate is moved from left to right. The pattern shown in the illustration is designed to provide a plain binary value which increases as the plate moves from left to right.

The main disadvantage of this arrangement is that it may require more than one bit to change simultaneously, e.g. consider what happens as the plate moves from position 7 (%0111) to 8 (%1000). This requires all four bits to change at the same time. For any real device, the actual bits sensed will alter at different instants as the plate moves from position 7 to 8. Hence between the correct readings of 7 and 8 we may find the encoder gives momentary readings of 15 (%1111), or 13 (%1101), or 12 (%1100), or 3 (%0011). In fact, as every bit has to change between 7 and 8, we

46

could momentarily get <u>any</u> number from 0 to 15 as the plate moves from one position to the other. A computer reading the sensed number at the wrong moment would therefore think the plate was leaping about in a frantic way as it moved from 7 to 8!

To avoid this problem we can replace the simple binary code with a new code system (i.e. change the pattern on the encoder plate) designed so that only <u>one</u> bit changes between adjacent locations. Two possible systems are the *Gray* code and a 'walking' code shown below:

#	Gray	Walking	#	Gray	Walking
00	%0000	%00000000	08	%1100	%11111111
01	%0001	%00000001	09	%1101	%11111110
02	%0011	%00000011	10	%1111	%11111100
03	%0010	%00000111	11	%1110	%11111000
04	%0110	%00001111	12	%1010	%11110000
05	%0111	%00011111	13	%1011	%11100000
06	%0101	%00111111	14	%1001	%11000000
07	%0100	%01111111	15	%1000	%10000000

Clearly the walking code uses redundancy to achieve its effect as it requires eight bits to cover the range 0 – 15. The Gray code is more interesting as it is simply a re-arrangement of the pure binary numbers from 0 – 15. The problem described above occurs because of imperfections in the way the plate and sensors are built. The errors produced <u>aren't</u> random. As we slowly move across the number boundaries the pattern of 'jumping about' is always the same for a given plate/sensor system. Errors of this kind are said to be *Systematic* since they depend upon fixed physical imperfections of the system we're using. The Gray code example shows that it is sometimes possible to devise a code which overcomes a specific *systematic* problem without <u>any</u> loss of efficiency. Dealing with random errors always requires a drop in efficiency. This is an important difference between errors produced by random noise and errors produced by repeatable, systematic effects.

The second example illustrates another weapon we can use to protect ourselves against mistaking received errors for reliable information. This technique is called *Soft Decision Making* and it depends upon being able to spot when received bits are 'suspect'. Combined with a block-checking code, this is a powerful way of reducing the effects of random noise.

To implement this technique we need to think about the transmission <u>method</u> rather than the code system. A simple example is electronic digital transmission along metal wires. Here bits can be lost due to momentary loss of contact (e.g. due to a rusty plug/socket somewhere) as well as random noise. This can produce *bursts* of errors where a series of bits are missed. The most direct method is to send, say, TTL voltage levels (between 0 and 1 V for '0' and between 3 and 5 V for '1'). A momentary loss of signal may produce either 0 V (received as '0') or allow a receiving TTL gate to float high (giving a received '1'). Hence, depending upon the receiver circuits used, a temporary loss of data looks like a string of '0's or '1's.

Various systems have been devised to avoid this. The most common is the transmission system called 'RS-232/432'. Here a positive current (typically about +3.5 mA) signals '1' and a negative one (−3.5 mA) signals a '0'. A momentary signal loss gives zero current which the receiver can respond to by tagging the appropriate bits as 'don't know'. It is worth noting that this method is essentially making use of <u>three</u> logic levels to send binary data; −3.5 mA='1', +3.5 mA='0', and 0 mA= 'don't know', although the transmitter is only attempting to send two of these.

In practice, this technique does have one potentially significant disadvantage which can be illustrated using the example of the RS232 logic levels. In the absence of any attempt to detect the 'don't know' condition the receiver could decide whether a '1' or '0' was being communicated by checking whether the received current was above or below 0 mA. Random noise would therefore have to change the current level by at least 3.5 mA in order to produce an error.

In order to be able to sense message interruptions the receiver must be designed so as to respond to some range of currents, $\pm I$, centred on 0 mA by deciding that the signal level is 'undefined'. Random noise now only has to alter the received current by an amount $(3.5 - |I|)$ mA, to make a bit appear unreliable. Similarly, the random noise only needs to produce a momentary current fluctuation of more than $|I|$ to make a momentary loss of signal as an apparently reliable '1' or '0'. This means that we can't avoid this problem by making I very small without giving up the ability to spot when data is failing to arrive. As a result, assigning an intermediate range of levels to mean 'undefined' leads to an increase in the frequency of errors produced by random noise. However, provided that the S/N ratio is high, this increase can be small enough to be an acceptable price for being able to sense momentary data losses.

Summary

You should now know that the effects of random errors can (usually!) be detected and corrected. In it's simplest forms this can be done using a method like 'tell me twice' or 'tell me three times' which repeat all the information. That parity bit generation methods and the use of block codes 'dilute' the amount of information repetition to provide a lower amount of protection but a higher transmission *Efficiency* (lower *Redundancy*) than simple 'tell me again'. You should also understand that the amount of protection from random errors depends upon the amount of redundancy since we require a given amount of extra 'illegal' symbols or bit-patterns in between the legal ones to be able to deal with random errors. You should also now know that the ability of a code system to detect and correct errors can be measured in terms of the code system's *Minimum Hamming Distance* value.

Finally, the example of the *Gray* code shows that <u>non-random</u> or *Systematic* errors can be corrected <u>without</u> the need for any extra bits or words — i.e. without any redundancy. The example of RS-232 shows that giving the receiver the ability to spot data losses, called *Soft Decision Making*, can be useful in dealing with *Bursts* of errors produced by problems like temporarily loss of contact with the transmitter.

Questions

1) A message is transmitted in the form of a series of digital bits. The signal is carried by a channel with a signal to noise ratio which means that each individual bit has a 0·9 chance of being received correctly. The message is 10,000 bits long. How many noise-produced random errors is a single copy of the message likely to contain when received? To try and reduce the effects of noise the message is sent using the *Tell Me Three Times* method. After error correction, how many undetected errors are likely to appear in the received message? [**1000. 280.**]

2) Explain what is mean by the terms *Parity Bit* and *Parity Checking*. A *Block Code* system groups 16 message bits at a time into a two-dimensional block in order to generate a set of parity bits. Draw a diagram of this process and explain how it enables single bits errors in a block to be detected and corrected. Explain why the presence of two bit errors in a block can be

detected but not always corrected using this system. What is the value of the transmitted signal's *Efficiency* including the parity bits? What is the value of the signal's *Redundancy*? [**Efficiency = 16/24. Redundancy = 0·333.**]

3) Draw a diagram of a *Linear Encoder* and use it to explain why the normal binary number sequence, %0000, %0001, %0010, etc., isn't a very suitable choice for the encoder pattern. Explain how either *Walking Code* or *Gray Code* can overcome the problem. Explain what advantage Gray Code has over Walking Code.

4) Explain the term *Soft Decision Making*. Give a brief explanation of how the *RS-232* data transmission system can indicate data losses during transmission.

Chapter 7

The Sampling Theorem

7.1 Fourier Transforms and signals of finite length

In the first few chapters we saw that the amount of information conveyed along a channel will depend upon its bandwidth (or response time), the maximum signal power, and the noise level. The way we estimated the effects of these was fairly rough. We now need to look at this fundamental question of a channel's information carrying *Capacity* more carefully. The amount of information contained in a message can be formally defined using the *Sampling Theorem*. The maximum information carrying capacity of a transmission channel can be defined using *Shannon's Equation*. Taken together, they provide the basis of the whole structure of Information Theory. Rather than tackle the Sampling Theorem or Shannon's Equation 'head on', it is useful to take a diversion and begin by considering the relationship between a time-varying signal and its *Frequency Spectrum*.

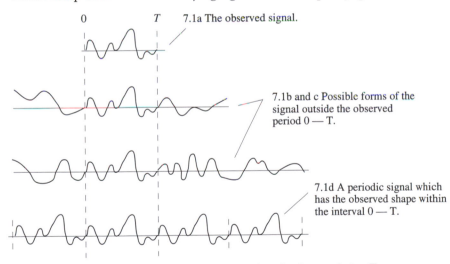

0   T   7.1a The observed signal.

7.1b and c Possible forms of the signal outside the observed period 0 — T.

7.1d A periodic signal which has the observed shape within the interval 0 — T.

Figure 7.1   A signal observed during the interval, 0 – T.

A message which requires an infinite time to finish isn't of any practical value. This is because we can't know what information it contains <u>until</u> it has all arrived! As a result, in practice we can only observe or deal with signals which have defined 'start' and 'stop' points. The fact that

51

information about a real signal or process can only cover a finite *duration* or *interval* has some important consequences.

Consider the situation illustrated in figure 7.1a. This shows how a particular analog signal is seen to vary over a time interval, $t = 0$ to $t = T$. (For simplicity we've 'switched on the clock' at the start of the observation. Note that this doesn't affect our conclusions.) Now the only message information we have is confined to the chosen time interval. Logically, therefore, we have to accept that if we had looked at the signal for at other times we might have seen any of the alternatives shown in figure 7.1b, c, etc. However, the limited information we have doesn't allow us to <u>know</u> what happened outside our observation. We can, of course, theorise about what we might have seen if we had observed what was happening at other times. Provided any hypothesis doesn't conflict with the information we possess it can be accepted for the purpose of argument.

The signal we have observed can be described by some specific function of time, $p\{t\}$, which is only known when $0 \leqslant t \leqslant T$. From the argument given above we can, in principle, imagine an infinite variety of theoretical functions, $p'\{t\}$, which are defined so that

$$p'\{t\} = p\{t\} \; ; \; 0 \leqslant t \leqslant T \qquad \qquad \text{...} (7.1)$$

but which allow $p'\{t\}$ to do whatever we like at other times.

Using functions like $p\{t\}$ or $p'\{t\}$ we can describe the behaviour of a signal in terms of its variations with time. An alternative method for describing a signal is to specify its *frequency spectrum* in terms of some suitable function, $S\{f\}$. We can then consider the signal level at any instant, t, as

$$p\{t\} = \sum a_n \text{Cos}\{2\pi f_n t + \phi_n\} \qquad \qquad \text{...} (7.2)$$

i.e. the signal is regarded as being composed of a series of contributions at a set of frequencies, f_n. The size of each contribution, a_n and its phase at $t = 0$, ϕ_n, being defined by the value of $S\{f\}$ at the appropriate frequency, f_n. (Note that this means that, in general, $S\{f\}$ must specify <u>two</u> values, an amplitude and a phase, hence it is most convenient to treat this as a function which produces a <u>complex</u> result.)

Clearly the *time domain* description, $p\{t\}$, of a signal and its *frequency domain* description, $S\{f\}$, must contain identical information if they are both to specify the same signal or message. The two functions must

52

therefore be linked in some way. Mathematically, this link can be made using the technique called *Fourier Transformation*.

Experience shows that it can be a mistake for a student to read more than one book which uses Fourier analysis! Comparing one text with another reveals a host of odd factors of 2, π, etc., which seem to pop up and disappear without any obvious reason. The most common result of this is to make most engineering and science students decide to avoid the topic whenever possible! Unfortunately, Fourier methods are very useful. Ignoring them is a bit like avoiding using saws when doing woodwork because you aren't sure which type of saw is best. Since this isn't a maths book we won't examine Fourier Transforms in detail, but it is worth making a few comments which may be helpful.

Firstly, we can see from equation 7.2 that to specify the effect of a given frequency component on a signal we need to have <u>two</u> values. In 7.2. these were an amplitude, a_n, and a phase, ϕ_n. We could, however, achieve the same effect in other ways. For example, we could define the same signal in terms of pairs of values, A_n and B_n, in an expression like

$$p\{t\} = \sum A_n \, \text{Cos}\{2\pi f_n t\} + B_n \, \text{Sin}\{2\pi f_n t\} \qquad \text{... (7.3)}$$

or we could use something like

$$p\{t\} = \sum \text{Real}\left[a_n \, \text{Exp}\{-j2\pi f_n t + \phi_n\}\right] \qquad \text{... (7.4)}$$

All of these are equivalent ways to achieve the same result, but they alter the form of the Fourier Transform expressions required to link the time and frequency domains.

Secondly, the form of the Fourier Transform expressions depends upon whether we are interested in knowing the <u>power</u> (or amplitude) of the signal or the total <u>energy</u> it conveys. This affects whether the expressions have to be multiplied by a factor proportional to $\frac{1}{T}$ since power = energy per unit time. Here we will use the type of expression given in 7.3 and consider the amplitude (e.g. the voltage) of the signals. This determines the details of the Fourier Integrals we'll use. In fact, we would come to the same conclusions using any of the other approaches.

Sines and cosines are an example of a set of *Orthogonal Functions*. The general topic of the properties of orthogonal functions is beyond the scope of this book. All we have to do is outline some of their basic properties which are relevant here. In general, a set of functions, $F_n\{z\}$, which satisfy the integral

$$\int_a^b F_n\{z\} F_m\{z\} \; dz \; = \; 0 \quad \text{when } n \neq m \qquad \qquad \dots (7.5)$$

are said to be 'orthogonal over the range a to b'. For the case of sine or cosine functions we can regard $F_n\{z\}$ and $F_m\{z\}$ as having two different angular frequencies, $\omega_n \equiv 2\pi f_n$, $\omega_m \equiv 2\pi f_m$. If we consult a book of integrals or a text on the properties of functions we can find that, provided $n \neq m$

$$\int_0^\pi \mathrm{Sin}\{mx\} \, \mathrm{Sin}\{nx\} \; dx \; = \; \int_0^\pi \mathrm{Cos}\{mx\} \, \mathrm{Cos}\{nx\} \; dx \; = \; 0 \quad \dots (7.6)$$

where m and n are integers. This is equivalent to saying

$$\int_0^T \mathrm{Sin}\{n\omega_0 t\} \, \mathrm{Sin}\{m\omega_0 t\} \; dt \; = \; \int_0^T \mathrm{Cos}\{n\omega_0 t\} \, \mathrm{Cos}\{m\omega_0 t\} \; dt \; = \; 0$$

$$\dots (7.7)$$

where $\omega_0 \equiv \pi / T$. We can interpret this as defining a 'fundamental frequency', $f_0 \equiv \frac{1}{2T}$, which can fit one half-cycle into the interval, T.

This orthogonal behaviour is very important for the usefulness of Fourier analysis. The reason for this can be understood by going back to the signal we considered at the start of this chapter. This is a signal, $p\{t\}$, whose value is known <u>only</u> during the interval, $0 \leqslant t \leqslant T$.

As we have seen, we can imagine a variety of functions, $p'\{t\}$, which are identical to $p\{t\}$ during this observed interval but behave however we wish at other times. Provided we always ensure that $p'\{t\} = p\{t\}$ during the signal interval every possible choice of $p'\{t\}$ provides us with <u>exactly the same information</u> (pattern) during this period as $p\{t\}$. All these possible choices are indistinguishable from one another if we only observe this finite interval. This gives us the freedom to choose <u>any</u> $p'\{t\}$ which is identical to $p\{t\}$ during the observed interval. We can therefore select one which is convenient for the purpose of analysing the signal. There is nothing to stop us from choosing a form for $p'\{t\}$ which is *Periodic* — i.e. one which repeats itself over and over again — with a period equal to the observed signal's interval, T. This assumption is convenient for the purposes of Fourier analysis. If we assume $p'\{t\}$ is periodic in this way it will take the form shown in figure 7.1d.

It should be clear that a signal which repeats itself in this way can <u>only</u> contain frequencies which are multiples of a fundamental frequency, $f_0 = 1/T$ (plus, perhaps, a non-zero d.c. level). This is because the presence of any other frequencies would mean each 'cycle' of the periodic function would differ from its neighbours. We can therefore say that the function must be of the form

$$p'\{t\} = \sum_{n=0}^{N} A_n \, \text{Cos}\left\{2\pi n f_0 t\right\} + B_n \, \text{Sin}\left\{2\pi n f_0 t\right\} \qquad \dots (7.8)$$

where N represents the highest frequency present and the $A_n \, B_m$ values determine the magnitude and phase of the n th frequency component of the signal. Note that this expression only contains a d.c. level ($n = 0$), a component at the fundamental frequency, f_0, and components at its *harmonic* frequencies, nf_0. (As $\text{Sin}\{0\} = 0$ and $\text{Cos}\{0\} = 1$ the d.c. level equals A_0. B_0 has no physical meaning.) Since this function is chosen so as to be indistinguishable from $p\{t\}$ during the observed period we can therefore say that $p\{t\}$ is indistinguishable from

$$p\{t\} = \sum_{n=0}^{N} A_n \, \text{Cos}\left\{2\pi n f_0 t\right\} + B_n \, \text{Sin}\left\{2\pi n f_0 t\right\} \qquad \dots (7.9)$$

during the observed period. The coefficients, A_n and B_n may be obtained from $p\{t\}$ using the Fourier integrals,

$$A_n = \frac{2}{T} \int_0^T p\{t\} \, \text{Cos}\left\{2\pi n f_0 t\right\} \, dt \qquad \dots (7.10)$$

$$B_n = \frac{2}{T} \int_0^T p\{t\} \, \text{Sin}\left\{2\pi n f_0 t\right\} \, dt \qquad \dots (7.11)$$

These expressions represent the Fourier Transform of the known signal, $p\{t\}$, and allow us to calculate the signal's frequency spectrum. (Expressions 7.10 and 7.11 can be seen to be true once we accept that 7.6 and 7.7 are correct. In effect, the above expressions let us 'pick out' the two coefficients we want from $p\{t\}$ at any chosen frequency, nf_0.)

From the above arguments it should be clear that we can freely convert information back and forth between the time domain and the frequency domain. Given this ability it must be true that the frequency spectrum contains the same information as the time-varying signal.

7.2 The Sampling Theorem and signal reconstruction

Any real signal will be transmitted along some form of channel which will have a finite bandwidth. As a result the received signal's spectrum cannot contain any frequencies above some maximum value, f_{max}. However, the spectrum obtained using the Fourier method described in the previous section will be characteristic of a signal which repeats after the interval, T. This means it can be described by a spectrum which only contain the frequencies, 0 (d.c.), f_0, $2f_0$, $3f_0$, ... Nf_0, where N is the largest integer which satisfies the inequality $Nf_0 \leqslant f_{max}$. As a consequence we can specify everything we know about the signal spectrum in terms of a d.c. level plus the amplitudes and phases of just N frequencies — i.e. all the information we have about the spectrum can be specified by just $2N+1$ numbers. Given that no information was lost when we calculated the spectrum it immediately follows that everything we know about the shape of the time domain signal pattern could also be specified by just $2N+1$ values.

For a signal whose duration is T this means that we can represent <u>all</u> of the signal information by measuring the signal level at $2N+1$ points equally spaced along the signal waveform. If we put the first point at the start of the message and the final one at its end this means that each sampled point will be at a distance $\frac{1}{2f_{max}}$ from its neighbours. This result is generally expressed in terms of the *Sampling Theorem* which can be stated as: '*If a continuous function contains no frequencies higher than f_{max} Hz it is completely determined by its value at a series of points less than $\frac{1}{2f_{max}}$ apart.*'

Consider a signal, $p\{t\}$, which is observed over the time interval, $0 \leqslant 0 \leqslant T$, and which we know cannot contain any frequencies above f_{max}. We can sample this signal to obtain a series of values, x_i, which represent the signal level at the instants, $t_i = \frac{iT}{K}$, where i is an integer in the range 0 to K. (This means there are $K+1$ samples.) Provided that $K \geqslant 2N$, where N is defined as above, we have satisfied the requirements of the Sampling Theorem. The samples will then contain all of the information present in the original signal and make up what is called a *Complete Record* of the original.

In fact, the above statement is a fairly 'weak' form of the sampling theorem. We can go on to a stricter form:

'If a continuous function only contains frequencies within a   bandwidth, *B* Hertz, it is completely determined by its value at a series of points spaced less than $1/(2B)$ seconds apart.'

This form of the sampling theorem can be seen to be true by considering a signal which doesn't contain any frequencies below some lower cut-off value, f_{min}. This means the values of A_n and B_n for low n (i.e. low values of $2\pi n f_0$) will all be zero. This limits the number of spectral components present in the signal just as the upper limit, f_{max}, means that there are no components above f_{max}. This situation is illustrated in figure 7.2.

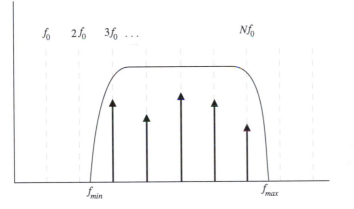

Figure 7.2   Spectrum of a band-limited signal of finite length.

From the above argument a signal of finite length, T, can be described by a spectrum which only contains frequencies, f_0, $2f_0$, ... Nf_0. If the signal is restricted to a given bandwidth, $B = f_{max} - f_{min}$, only those components inside the band have non-zero values. Hence we only need to specify the A_n and B_n values for those components to completely define the signal. The minimum required *sampling rate* therefore depends upon the bandwidth, not the maximum frequency. (Although in cases where the signal has components down to d.c. the two are essentially the same.)

The sampling theorem is of vital importance when processing information as it means that we can take a series of samples of a continuously varying signal and use those values to represent the entire signal without <u>any</u> loss of the available information. These samples can later be used to reconstruct <u>all</u> of the details of the original signal — even recovering details of the actual signal pattern 'in between' the sampled moments. To demonstrate this we can show how the original waveform can be 'reconstructed' from a complete set of samples.

The approach used in the previous section to calculate a signal's spectrum depends upon being able to integrate a continuous analytical function. Now, however, we need to deal with a set of sampled values instead of a continuous function. The integrals must be replaced by equivalent summations. These expressions allow us to calculate a frequency spectrum (i.e. the appropriate set of A_n and B_n values) from the samples which contain all of the signal information. The most obvious technique is to proceed in two steps. Firstly, to take the sample values, x_i, and calculate the signal's spectrum. Given a series of samples we must use the series expressions

$$A_n = \frac{2}{K} \sum_{i=0}^{K} x_i \, Cos\{2\pi nf_0 t_i\} \qquad B_n = \frac{2}{K} \sum_{i=0}^{K} x_i \, Sin\{2\pi nf_0 t_i\} \dots (7.12)$$

to calculate the relevant spectrum values. These are essentially the equivalent of the integrals, 7.10 and 7.11, which we would use to compute the spectrum of a continuous function. The second step of this approach is to use the resulting A_n and B_n values in the expression

$$x\{t\} = \sum_{n=0}^{N} A_n \, Cos\{2\pi nf_0 t\} + B_n \, Sin\{2\pi nf_0 t\} \qquad \dots (7.13)$$

to compute the signal level at <u>any</u> time, t, during the observed period. In effect, this second step is simply a restatement of the result shown in expression 7.9. Although this method works, it is computationally intensive and indirect. This is because it requires us to perform a whole series of numerical summations to determine the spectrum, followed by another summation for each $x\{t\}$ we wish to determine. A more straightforward method can be employed, based upon combining these operations. Expressions 7.12 and 7.13 can be combined to produce

$$x\{t\} =$$

$$\sum_{n=0}^{N} \frac{2}{K} \sum_{i=0}^{K} x_i \left[Cos\{2\pi nf_0 t_i\} \, Cos\{2\pi nf_0 t\} + Sin\{2\pi nf_0 t_i\} \, Sin\{2\pi nf_0 t\} \right]$$

$$\dots (7.14)$$

which, by a fairly involved process of algebraic manipulation, may be simplified into the form

$$x\{t\} = \sum_{i=0}^{K} x_i \text{ Sinc}\left\{\frac{\pi(t - t_i)}{\Delta t}\right\} \qquad \dots (7.15)$$

where the Sinc function can be defined as

$$\text{Sinc}\{z\} \equiv \frac{\text{Sin}\{z\}}{z} \qquad \dots (7.16)$$

and $\Delta t = T / K$ is the time interval between successive samples.

Given a set of samples, x_i, taken at the instants, t_i, we can now use expression 7.15 to calculate what the signal level would have been at <u>any</u> time, t, during the sampled signal interval.

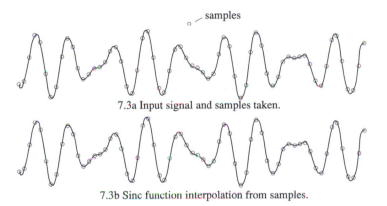

7.3a Input signal and samples taken.

7.3b Sinc function interpolation from samples.

Figure 7.3 Signal reconstruction from a series of sampled values.

Clearly, by using this approach we can calculate the signal value at any instant by performing a single summation over the sampled values. This method is therefore rather easier (and less prone to computational errors!) than the obvious technique. Figure 7.2 was produced by a **BBC Basic** program to demonstrate how easily this method can be used.

Although the explanation given here for the derivation of expression 7.15 is based upon the use of a Fourier technique, the result is a completely general one. Expression 7.15 can be used to 'interpolate' <u>any</u> given set of sampled values. The only requirement is that the samples have been obtained in accordance with the Sampling Theorem and that they do, indeed, form a complete record. It is important to realise that, under these circumstances, the recovered waveform is not a 'guess' but a reliable reconstruction of what we would have observed if the original signal had been measured at these other moments.

Summary

You should now be aware that the information carried by a signal can be defined either in terms of its *Time Domain* pattern or its *Frequency Domain* spectrum. You should also know that the amount of information in a continuous analog signal can be specified by a finite number of values. This result is summarised by the *Sampling Theorem* which states that we can collect <u>all</u> the information in a signal by sampling at a rate $2B$, where B is the signal bandwidth. Given this information we can, therefore, reconstruct the actual shape of the original continuous signal at any instant 'in between' the sampled instants. It should also be clear that this reconstruction is not a guess but a true reconstruction.

Questions

1) A single microphone is used to make an analog recording of a song 3 minutes long. The microphone only responds to signals in the 10 Hz to 18 kHz frequency range. How many digital samples are required to convert all the song's information into a complete digital record? [**6·47 million.**]

2) A complex signal is digitally recorded for 1 minute. The recorded information is then used to work out the spectrum of the observed spectrum. What will be the value of the frequency resolution of the spectrum we obtain? [**1/60th of a Hertz.**]

Chapter 8

The information carrying capacity of a channel

8.1 Signals look like noise!

One of the most important practical questions which arises when we are designing and using an information transmission or processing system is, 'What is the *Capacity* of this system? — i.e. How much information can it transmit or process in a given time?' We formed a rough idea of how to answer this question in an earlier chapter. We can now go on to obtain a more well defined answer by deriving *Shannon's Equation*. This equation allows us to precisely determine the information carrying capacity of <u>any</u> signal channel.

Consider a signal which is being efficiently communicated (i.e. no redundancy) in the form of a time-dependent analog voltage, $V\{t\}$. The pattern of voltage variations during a specific time interval, T, allows a receiver to identify which one of a possible set of messages has actually been sent. At any two moments, t_1 and t_2, during a message the voltage will be $V\{t_1\}$ and $V\{t_2\}$.

Using the idea of intersymbol influence we can say that — since there is no redundancy — the values of $V\{t_1\}$ and $V\{t_2\}$ will appear to be independent of one another provided that they're far enough apart (i.e. $|t_1 - t_2| > \frac{1}{2B}$) to be worth sampling separately. In effect, we can't tell what one of the values is just from knowing the other. Of course, for any <u>specific</u> message, both $V\{t_1\}$ and $V\{t_2\}$ are determined in advance by the content of that particular message. But the receiver can't know which of all the possible messages has arrived until it <u>has</u> arrived. If the receiver <u>did</u> know in advance which voltage pattern was to be transmitted then the message itself wouldn't provide any new information! That is because the receiver wouldn't know any more after its arrival than before. This leads us to the remarkable conclusion that a signal which is efficiently communicating information will vary from moment to moment in an unpredictable, apparently random, manner. An efficient signal looks very much like random noise!

61

This, of course, is why random noise can produce errors in a received message. The statistical properties of an efficiently signalled message are similar to those of random noise. If the signal and noise were obviously different the receiver could easily separate the noise from the signal and avoid making any errors.

To detect and correct errors we therefore have to make the real signal less 'noise-like'. This is what we're doing when we use parity bits to add redundancy to a signal. The redundancy produces predictable relationships between different sections of the signal pattern. Although this reduces the system's information carrying efficiency it helps us distinguish signal details from random noise. Here, however, we're interested in discovering the maximum possible information carrying capacity of a system. So we have to avoid any redundancy and allow the signal to have the 'unpredictable' qualities which make it statistically similar to random noise.

The amount of noise present in a given system can be represented in terms of its mean noise power

$$N = V_N^2/R \qquad \ldots (8.1)$$

where R is the characteristic impedance of the channel or system and V_N is the rms noise voltage. In a similar manner we can represent a typical message in terms of its average signal power

$$S = V_S^2/R \qquad \ldots (8.2)$$

where V_S is the signal's rms voltage.

A real signal must have a finite power. Hence for a given set of possible messages there must be some maximum possible power level. This means that the rms signal voltage is limited to some range. It also means that the instantaneous signal voltage must be limited and can't be beyond some specific range, V_S'. A similar argument must also be true for noise. Since we are assuming that the signal system is efficient we can expect the signal and noise to have similar statistical properties. This implies that if we watched the signal or noise for a long while we'd find that their level fluctuations had the same peak/rms voltage ratio. We can therefore say that, during a typical message, the noise voltage fluctuations will be confined to some range

$$\pm V_N' = \pm \eta V_N \qquad \ldots (8.3)$$

where the *form factor*, η, (ratio of peak to rms levels) can be defined from the signal's properties as

62

$$\eta \equiv \frac{V'_S}{V_S} \qquad \ldots (8.4)$$

When transmitting signals in the presence of noise we should try to ensure that S is as large as possible so as to minimise the effects of the noise. We can therefore expect that an efficient information transmission system will ensure that, for every typical message, S is almost equal to some maximum value, P_{max}. This implies that in such a system, most messages will have a similar power level. Ideally, every message should have the same, maximum possible, power level. In fact we can turn this argument on its head and say that <u>only</u> messages with mean powers similar to this maximum are 'typical'. Those which have much lower powers are unusual — i.e. rare.

8.2 Shannon's equation

The signal and noise are *Uncorrelated* — that is, they are not related in any way which would let us predict one of them from the other. The total power obtained, P_T, when combining these uncorrelated, apparently randomly varying quantities is given by

$$P_T = S + N \qquad \ldots (8.5)$$

i.e. the typical combined rms voltage, V_T, will be such that

$$V_T^2 = V_S^2 + V_N^2 \qquad \ldots (8.6)$$

Since the signal and noise are statistically similar their combination will have the same form factor value as the signal or noise taken by itself. We can therefore expect that the combined signal and noise will generally be confined to a voltage range $\pm \eta V_T$.

Consider now dividing this range into 2^b bands of equal size. (i.e. each of these bands will cover $\Delta V = 2\eta V_T / 2^b$.) To provide a different label for each band we require 2^b symbols or numbers. We can then always indicate which ΔV band the voltage level occupies at any moment in terms of a unique b-bit binary number. In effect, this process is another way of describing what happens when we take digital samples with a b-bit analog to digital convertor working over a total range $2V_T$.

There is no real point in choosing a value for b which is so large that ΔV is smaller than $2\eta V_N$. This is because the noise will simply tend to randomise the actual voltage by this amount, making any extra bits meaningless. As a result the maximum number of bits of information we can obtain regarding the level at any moment will given by

$$2^b = \frac{V_T}{V_N} \qquad \qquad \text{... (8.7)}$$

i.e.

$$2^b = \sqrt{\frac{V_T^2}{V_N^2}} = \sqrt{\left(\frac{V_N^2}{V_N^2}\right) + \left(\frac{V_S^2}{V_N^2}\right)} = \sqrt{1 + (S/N)} \qquad \text{... (8.8)}$$

which can be rearranged to produce

$$b = \text{Log}_2\left\{\left(1 + \frac{S}{N}\right)^{1/2}\right\} \qquad \qquad \text{... (8.9)}$$

If we make M, b-bit measurements of the level in a time, T, then the total number of bits of information collected will be

$$H = Mb = M.\,\text{Log}_2\left\{\left(1 + \frac{S}{N}\right)^{1/2}\right\} \qquad \qquad \text{... (8.10)}$$

This means the information transmission *rate*, I, bits per unit time, will be

$$I = \left(\frac{M}{T}\right)\text{Log}_2\left\{\left(1 + \frac{S}{N}\right)^{1/2}\right\} \qquad \qquad \text{... (8.11)}$$

From the Sampling Theorem we can say that, for a channel of bandwidth, B, the highest practical sampling rate, M/T, at which we can make independent measurements or samples of a signal will be

$$\frac{M}{T} = 2B \qquad \qquad \text{... (8.12)}$$

Combining expressions 8.11 and 8.12 we can therefore conclude that the maximum information transmission rate, C, will be

$$C = 2B\,\text{Log}_2\left\{\left(1 + \frac{S}{N}\right)^{1/2}\right\} = B\,\text{Log}_2\left\{1 + \frac{S}{N}\right\} \qquad \text{... (8.13)}$$

This expression represents the maximum possible rate of information transmission through a given channel or system. It provides a mathematical proof of what we deduced in the first few chapters. The maximum rate at which we can transmit information is set by the bandwidth, the signal level, and the noise level. C is therefore called the channel's information carrying *Capacity*. Expression 8.13 is called *Shannon's Equation* after the first person to derive it.

8.3 Choosing an efficient transmission system

In many situations we are given a physical channel for information transmission (a set of wires and amplifiers, radio beams, or whatever) and have to decide how we can use it most efficiently. This means we have to assess how well various information transmission systems would make use of the available channel. To see how this is done we can compare transmitting information in two possible forms — as an analog voltage and a serial binary data stream — and decide which would make the best use of a given channel.

When doing this it should be remembered that there are a large variety of ways in which information can be represented. This comparison only tells us which out of the two we've considered is better. If we really did want to find the 'best possible' we might have to compare quite a few other methods. For the sake of comparison we will assume that the signal power at our disposal is the same regardless of whether we choose a digital or an analog form for the signal. It should be noted, however, that this isn't always the case and that any variations in available signal power with signal form will naturally affect the relative merits of the choices.

Noise may be caused by various physical processes, some of which are under our control to some extent. Here, for simplicity, we will assume that the only significant noise in the channel is due to unavoidable thermal noise. Under these conditions the noise power will be

$$N = kTB \qquad \qquad ...(8.14)$$

where T is the physical temperature of the system, and k is Boltzmann's constant.

Thermal noise has a 'white' spectrum — i.e. the noise power spectral density is the same at all frequencies. Many of the other physical processes which generate noise also exhibit white spectra. As a consequence we can often describe the overall noise level of a real system in terms of a *Noise Temperature, T,* which is linked to the observed total noise by expression 8.14. The concept of a noise temperature is a convenient one and is used in many practical situations. Its important to remember, however, that a noisy system may have a noise temperature of, say, one million Kelvins, yet have a physical temperature of no more than 20 °C! The noise temperature isn't the same thing as the 'real' temperature. A very noisy amplifier doesn't have to glow in the dark or emit X-rays!

Most real signals begin in an analog form so we can start by considering an analog signal which we wish to transmit. The highest frequency component in this signal is at a frequency, W Hz. The Sampling Theorem tells us that we would therefore have to take at least $2W$ samples per second to convert all the signal information into another form. If we

65

choose to transmit the signal in analog form we can place a low-pass filter in front of the receiver which rejects any frequencies above W. This filter will not stop any of the wanted signal from being received, but rejects any noise power at frequencies above W. Under these conditions the effective channel bandwidth will be equal to W and the received noise power, N, will be equal to kTW. Using Shannon's equation we can say that the effective capacity of this analog channel will be

$$C_{analog} = W \, \mathrm{Log}_2 \left\{ 1 + \frac{S}{kTW} \right\} \qquad \ldots (8.15)$$

In order to communicate the same information as a serial string of digital values we have to be able to transmit two samples of m bits each during the time required for one cycle at the frequency, W— i.e. we have to transmit $2mW$ bits per second. The frequencies present in a digitised version of a signal will depend upon the details of the pattern of '1's and '0's. The highest frequency will, however, be required when we alternate '1's and '0's. When this happens each pair of '1's and '0's will look like the high and low halves of a signal whose frequency is mW (<u>not</u> $2mW$). Hence the digital signal will require a channel bandwidth of mW to carry information at the same rate as the analog version.

Various misconceptions have arisen around the question of the bandwidth required to send a serial digital signal. The most common of these amongst students (and a few of their teachers!) are:-

i) *'Since you are sending 2mW bits per second, the required digital bandwidth is 2mW.'*

ii) *'Since digital signals are like squarewaves, you have to provide enough bandwidth to keep the 'edges square' so you can tell they're bits, not sinewaves.'*

<u>Neither of the above statements are true</u>. The required signal bandwidth is determined by how quickly we have to be able to switch level from '1' to '0' and vice versa. The digital receiver doesn't have to see 'square' signals, all it has to do is decide which of the two possible levels is being presented during the time allotted for any specific bit.

In order to allow all the digital signal into the receiver whilst rejecting 'out of band' noise we must now employ a noise-rejecting filter in front of the receiver which only rejects frequencies above mW. The effective capacity of this digital channel will then be

$$C_{digital} = mW \, \mathrm{Log}_2 \left\{ 1 + \frac{S}{kTmW} \right\} \qquad \ldots (8.16)$$

This shows the capacity of the channel at our disposal <u>if</u> we can set the bandwidth to the value required to send the data in digital serial form.

Note that this is <u>not</u> the actual rate at which we wish to send data! The digital data rate is

$$I = 2mW \qquad \qquad \text{...}(8.17)$$

It will only be possible to transmit the data in digital form <u>if</u> we can satisfy two conditions:

  i) The channel must actually be able to transmit frequencies up to mW.

  ii) The capacity of the channel must be greater or equal to I.

The digital form of signal will only communicate information at a higher rate than the analog form if

$$I > C_{analog} \qquad \qquad \text{...}(8.18)$$

so there is no point in digitising the signal for transmission unless this inequality is true. The number of bits per sample, m, must therefore be such that

$$m > \left(\frac{1}{2}\right) \text{Log}_2 \left\{ 1 + \frac{S}{kTW} \right\} \qquad \qquad \text{...}(8.19)$$

Otherwise the precision of the digital samples will be worse than the uncertainty introduced into an analog version of the signal by the channel noise. As a result, if the digital system is to be better than the analog one, the number of bits per sample must satisfy 8.19. (Note that this also means the initial signal has to have a S/N ratio good enough to make it worthwhile taking m bits per sample!)

Unfortunately, we can't just choose a value for m which is as large as we would always wish. This is because the data rate, I, cannot exceed the digital channel capacity, $C_{digital}$. From 8.16 and 8.17 this is equivalent to requiring that

$$2mW \leqslant mW \, \text{Log}_2 \left\{ 1 + \frac{S}{kTmW} \right\} \qquad \qquad \text{...}(8.20)$$

i.e.

$$m \leqslant \frac{S}{3kTW} \qquad \qquad \text{...}(8.21)$$

We can therefore conclude that a digitised form of signal will convey more information than an analog form over the available channel <u>if</u> we can choose a value for m which simultaneously satisfies conditions 8.19 and 8.21, and the available channel can carry a bandwidth, mW. If we can't satisfy these requirements the digital signalling system will be poorer than the analog one.

8.4 Noise, quantisation, and dither

An unavoidable feature of digital systems is that there must always be a finite number of bits per sample. This affects the way details of a signal will be transmitted.

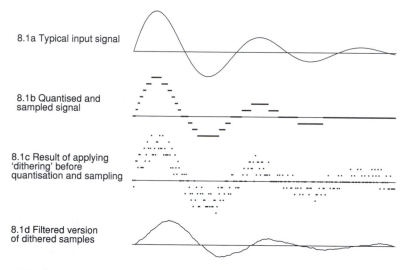

8.1a Typical input signal

8.1b Quantised and sampled signal

8.1c Result of applying 'dithering' before quantisation and sampling

8.1d Filtered version of dithered samples

Figure 8.1 The use of 'dithering' to overcome quantisation distortion.

Figure 8.1a represents a typical example of an input analog signal. In this case the signal was obtained from the function $\mathrm{Sin}\left\{ax\right\}\,\mathrm{Exp}\left\{-bx\right\}$ — i.e. an exponentially decaying sinewave. Figure 8.1b shows the effect of converting this into a stream of 4-bit digital samples and communicating these samples to a receiver which restores the signal into an analog form. Clearly, figures 8.1a and 8.1b are <u>not</u> identical! The received signal (figure 8.1b) has obviously been *Distorted* during transmission and is no longer a precise representation of the input. This distortion arises because the communication system only has $2^4 = 16$ available code symbols or levels to represent the variations of the input signal. The output of the system is said to be *Quantised*. It can only produce one of the sixteen available possible levels at any instant. The difference between adjacent levels is called the *Quantisation Interval*. Any smooth changes in the input become converted into a 'staircase' output whose steps are one quantisation interval high.

This form of distortion is particularly awkward when we are interested in the small details of a signal. Consider, for example, the low-amplitude fluctuations of the 'tail' of the signal shown in figure 8.1a. These variations are totally absent from the received signal shown in figure 8.1b. This is because the digitising system uses the same symbol for all of the levels of this small tail. As a result we can expect that any details of the signal which involve level changes smaller than a quantisation interval may be entirely lost during transmission.

At first sight these quantisation effects seem unavoidable. We can reduce the severity of the quantisation distortion by increasing the number of bits per sample. In our 4-bit example the quantisation interval is $1/2^4$ th of the total range (6·25%). If were to replace this with a Compact Disc standard system using 16-bit samples the quantisation interval would be reduced to $1/2^{16}$th (0.0015%). This reduces the staircase effect, but doesn't banish it altogether. As a result, small signal details will, it seems, always be lost. Fortunately, there is a way of dealing with this problem. We can add some random noise to the signal before it is sampled. Noise which has been deliberately added in this way to a signal before sampling is called *Dither*.

Figure 8.1c shows the kind of received signal we will obtain if some noise is added to the initial signal before sampling. This noise has the effect of superimposing a random variation onto the staircase distortion. Figure 8.1d shows the effect of passing the output shown in figure 8.1c through a filter which smooths away the higher frequencies. This essentially produces a 'moving average' of the received signal plus noise. This filtering action can be carried out by passing the output from the receiver's digital-to-analog convertor through a low-pass analog filter (e.g. a simple *RC* time constant). Alternatively, filtering can be carried out by performing some equivalent calculations upon the received digital values before reconversion into an analog output. This 'numerical' approach was adopted for the example shown in figure 8.1.

Comparing figures 8.1d and 8.1b we can see that the combination of input dithering and output filtering can remove the quantisation staircase. We may therefore conclude that *Dithering* provides a way to overcome this form of distortion. It can also (as shown) allow the system to communicate signal details such as the small 'tail' of the waveform which are smaller than the quantisation. In reality any input signal will already contain some random noise, however small. In principle therefore we don't need to add any extra noise if, instead, we can employ an analog-to-digital convertor (ADC) which produces enough bits per sample to ensure that the quantisation interval is less than the pre-existing noise level. All that matters is that the signal presented to the ADC varies randomly by an amount greater than the quantisation interval. In principle, the amount of information communicated is not significantly altered by using dithering. However, the form of information loss changes from a 'hard' staircase distortion loss to a 'gentle' superimposed random noise which is often more acceptable – for example, in audio systems, where the human ear is less annoyed by random noise than periodic distortions. The ability of dithered systems to respond to tiny signals well below the quantisation level is also useful in many circumstances. Hence dither is widely used when signals are digitised.

From a practical point of view using random noise in this way is quite useful. Most of the time engineers and scientists want to reduce the noise level in order to make more accurate measurements. Noise is usually regarded as an enemy by information engineers. However when digitising analog signals we <u>want</u> a given amount of noise to avoid quantisation effects. The noise allows us to detect small signal details by averaging over a number of samples. Without the noise these details would be lost since small changes in the input signal level would leave the output unchanged.

In fact, the use of dither noise in this way is a special case of a more general rule. Consider as an example a situation where you are using a 3-digit *Digital VoltMeter* (DVM) to measure a d.c. voltage. In the absence of any noise you get a steady reading, something like 1·29 V, say. No matter how long you stare at the DVM, the value remains the same. In this situation, if you want a more accurate measurement you may have to get a more expensive DVM which shows more digits! However, if there is a large enough amount of random noise superimposed on the d.c. you'll see the DVM reading vary from time to time. If you now regularly note the DVM reading you'll get some sequence like, 1·29, 1·28, 1·29, 1·27, 1·26, 1·29, etc... Having collected enough measurements you can now add up all the readings and take their average. This can provide a <u>more</u> accurate result than the steady 1·29 V you'd get from a steady level in the absence of any noise.

We'll be looking at the use of *Signal Averaging* in more detail in a later chapter. Here we need only note that, for averaging to work, we must have a random level fluctuation which is at least a little larger than the quantisation interval. In the case of the 3-digit DVM the quantisation level is the smallest voltage change which alters the reading — i.e. 0·01 Volts in this example. In the case of the 4-bit analog to digital/digital to analog system considered earlier it is $1/2^4$ of the total range. Although the details of the two examples differ, the basic usefulness of dither and averaging remains the same.

Summary

You should now know that an efficient (i.e. no redundancy or repetition) signal provides information because its form is unpredictable in advance. This means that its statistical properties are the same as random noise. You should also now know how to use *Shannon's Equation* to determine the information carrying capacity of a channel and decide whether a digital or analog system makes the best use of a given channel. You should now know how quantisation distortion arises. It should also be clear that a properly dithered digital information system can provide an output signal which looks just like an analog 'signal plus noise' output without any signs of quantisation.

Questions

1) Explain what we mean by the *Capacity* of an information carrying channel. A channel carries a signal whose maximum possible peak-to-peak voltage is $V_S = 1$ V and has a peak-to-peak noise voltage, $V_N = 0.001$ V. The bandwidth of the channel is $B = 10$ kHz. Derive *Shannon's Equation* and use it to calculate the value of the channel's capacity. [**199,314 bits/ second.**]

2) Explain what we mean by the *Noise Temperature* of a system. A channel has a bandwidth of 100 kHz and is used to carry a serial digital signal. The signal is produced by an 8-bit analog to digital convertor fed by an analog input. How many samples per second can the system carry? The signal power level is 1 μW. What is the highest noise temperature value which would still let the system carry the digital signal successfully? [**25,000 samples/second. 2.4×10^{11} K.**]

3) Using the same channel as above, what is the highest noise temperature which would be acceptable if the channel were used to carry the information in its original <u>analog</u> form? [**8.8×10^7 K.**]

4) Explain what we mean by the term *Dither* and say how it can be used to overcome *Quantisation Distortion* effects.

71

Chapter 9

The CD player as an information channel

9.1 The CD as an information channel

The next few chapters use the example of the CD audio system to show some of the basic properties of instruments used to gather and process information. CD has been selected for various reasons. It provides an excellent example of many digital data processing methods and allows us to explore the relationship between signals held in equivalent analog and digital forms. Both the source information gathering (i.e. the recording studio, etc.) and the information replay system (the CD player) can be used to illustrate a variety of highly effective measurement and information processing techniques. The CD system can also be simultaneously regarded as:

  i) A measurement system, collecting audio information.

  ii) A signal processing system.

  iii) An information communication channel/storage system.

The decision to choose CD for close examination is also based upon the thought that most science and engineering students will have a CD player and will be interested in understanding how it works.

Usually, texts on information theory tend to concentrate on systems where an information source and a receiver are directly connected by some channel. Information is then communicated through the channel in *Real Time*. Arrangements which store information for recovery at a later time can also be considered as communication systems. In general, the ideas and techniques of information theory can be applied equally well to both real time and stored or 'delayed' messages. The disc recording process then becomes an information transmitter or source. The CD player is a form of information 'receiver', and the disc itself is an information 'channel'.

When designing or choosing any information transmission system we must start by defining the properties of the signals we wish it to carry. The Compact Disc has to communicate two channels of *Audio* information, recorded in a form which can be used to reconstruct a *Stereo* soundfield. As with most human forms of communication the actual requirements would vary from one case to another. For example, some people can hear sounds at frequencies well above 20 kHz whereas others cannot hear 14 kHz. As a result there is not an 'obviously correct' choice for the required signal bandwidth. We will not consider whether a 'better' specification for the CD system would have produced an audible improvement. We shall simply examine the system as it has been implemented.

The CD system has been based upon the assumption that high fidelity sound reproduction requires a uniform frequency response from below 10 Hz to above 20 kHz and a dynamic range of more than 90 dB. This led to the decision to sample each of the stereo channels (left and right) 44,100 times per second, and to take 16-bit digital samples.

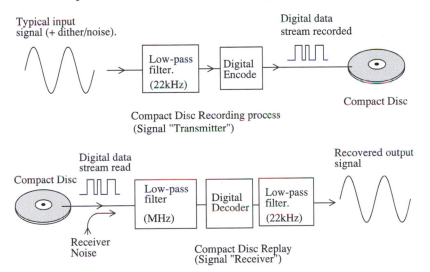

Figure 9.1 Compact Disc as a data communications channel.

Using 16-bit words, the ratio of the largest possible signal (which does not go out of range) to the quantisation interval is $1:2^{16} = 1:65,536$. This voltage ratio is equivalent to a power ratio of 96·3 dB, so we can expect the dynamic range of the CD system to be of this order. The input to a CD digital recording system is normally *dithered* in order to suppress quantisation distortion. From the sampling theorem we can expect that the chosen sampling rate will allow frequencies up to $44.1/2 = 22·05$ kHz to be recorded and replayed.

If we could be certain that the input signal would never contain any

73

components at frequencies above 22·05 kHz we could simply amplify the initial stereo signals to an appropriate level and present them to a pair of analog to digital convertors (ADCs) to obtain the required stream of digital samples. Unfortunately human speech and music does occasionally contain components at nominally inaudible frequencies well above 20 kHz. If these are allowed to reach the ADCs they will produce a particularly severe form of anharmonic signal distortion called *Aliasing*. This problem can be understood by considering the situation illustrated in figure 9.2.

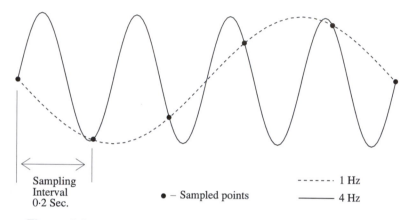

Sampling
Interval
0·2 Sec.

● – Sampled points

------- 1 Hz

——— 4 Hz

Figure 9·2 Demonstration that the same set of sampled values can be produced by different input signals of distinct frequencies.

For the sake of example, the illustration shows the results of sampling an input 4 Hz sinewave every 0·2 seconds (i.e. the *sampling rate* is 5 Hz). Looking at the figure we can see that an input 1 Hz sinewave could have produced exactly the same sample values as the 4 Hz wave. When the samples are presented to a Digital to Analog Convertor (DAC) for reconversion back into an analog waveform the result will be an output which looks identical to what we would get if the original input had been at 1 Hz. The 4 Hz input is said to be an *alias* of a 1 Hz input since it produces exactly the same output.

This aliasing effect gives us a serious problem if the input signal is allowed to contain frequency components at both 1 Hz and 4 Hz. The problem arises because we have not obeyed the sampling theorem. In order to pass 1 Hz – 4 Hz the input signal bandwidth must be at least 4 – 1 = 3 Hz. To satisfy the sampling theorem we would therefore have to take <u>at least</u> 6 samples per second — i.e. use a sampling rate of 6 Hz. The 5 Hz rate being used simply isn't enough to provide all the information needed to

74

recognise whether the input was at 1 Hz or 4 Hz. Unless we take steps to avoid it, aliasing can, therefore, produce significant signal *Distortions* causing the output to be very different from the input.

In fact the situation is even worse than the above implies. This is because the same set of samples could have been produced by an input signal at 6 Hz, or 9 Hz, or 11 Hz, or... When using a sampling rate, f_r, a frequency component at <u>any</u> frequency,,

$$f' = \frac{nf_r}{2} \pm f \qquad \qquad ...(9.1)$$

where n is any integer will produce a set of sampled values which are indistinguishable from those which would be produced by the signal frequency, f.

A CD player uses a sampling rate of 44.1 kHz, not 5 Hz, so it isn't likely to have trouble telling the difference between 1 Hz and 5 Hz! However, it <u>will</u> have problems if it is presented with input signal frequencies equal to or above 22·05 kHz. In order to avoid this possible source of signal distortion it is vital to use a pair of low-pass filters and stop frequencies $\geqslant$ 22·05 kHz from reaching the ADCs used to encode the CD audio signals.

9.2 The CD encoding process

For the CD system we can define, $m = 16$, to be the number of bits per sample and, $f_r = 44,100$, to be the number of samples per second taken of <u>each</u> of the two stereo signals. The required information transmission rate, I, is therefore

$$I = 2f_r m = 1,411,200 \quad \text{bits/sec} \qquad ...(9.2)$$

where the 2 is required because we wish to send stereo information. We therefore require a channel whose capacity, C, is at least 1·4112 Mbits/s. To send this information as a serial binary data stream we need a channel bandwidth, $B \geqslant I/2$. To minimise the effects of noise without losing any signal we should employ another low-pass filter to restrict the bandwidth entering the receiver (the CD's decoder circuits) to

$$B = \frac{I}{2} = 705 \cdot 6 \quad \text{kHz} \qquad ...(9.3)$$

(A bigger bandwidth passes more noise. A smaller one cuts off some signal.)

We can now apply Shannon's equation to say that, for a channel noise level of kT per unit of bandwidth, the signal power, S, needed for information to be successfully communicated will be such that

$$B \, \text{Log}_2 \left\{ 1 + \frac{S}{kTB} \right\} \geqslant I \qquad \qquad \text{...(9.4)}$$

Combining 9.3 and 9.4 we can say that the required signal power will be

$$S \geqslant 3kTB \qquad \qquad \text{...(9.5)}$$

Note that S increases with B. This is because the noise power entering the receiver increases with the bandwidth.

Consider now what would happen if we tried to employ the same channel to transmit just one of the pair of stereo signals in analog form. For the sake of comparison we can assume that the maximum signal power available for analog transmission is the same as the amount, $3kTB$, which would be just enough for the digital system to work. The receiver filter could be altered to restrict the received signal bandwidth to a value, $W = 22 \cdot 05$ kHz. This would then produce a signal to noise ratio of

$$\left(\frac{S}{N} \right)_{analog} = \frac{3kTB}{kTW} \qquad \qquad \text{...(9.6)}$$

i.e.

$$\left(\frac{S}{N} \right)_{analog} = \frac{3B}{W} \qquad \qquad \text{...(9.7)}$$

Given $B = 705.6$ kHz and $W = 22.05$ kHz, the analog system will provide a maximum S/N ratio (i.e. a dynamic range) of $3B/W = 96$ (19.8dB). The CD system employs 16-bit samples and can provide a dynamic range of about 95dB — i.e. 75dB better! This comparison shows that an analog signal can get through a smaller channel bandwidth, but it is much more susceptible to noise than a digital signal.

On the basis of the figures given above we can expect that a CD lasting 60 minutes will have to store $1 \cdot 4112 \times 60 \times 60 = 5,080$ Mbits. In fact, CDs employ a powerful error detection and correction system — i.e. the codes used include some redundancy. Although the amount of information on a 60 minute CD remains around 5 Gbits, the number of recorded bits is much greater. This means that the rate at which data bits are read from the disc (and the receiver's channel bandwidth, B) must be somewhat higher than we've assumed.

The encoding scheme employed for CD is quite complex. Fortunately we only need to consider its main elements to appreciate how the basic concepts of information theory have been applied. The explanation given here is based upon information provided by *Philips* (who developed the CD system along with Sony) in a special issue of the *Philips Technical Review* (Vol. 40(6) 1982).

Figure 9.3 represents the CD encoding/recording system. The input data is initially sampled in the form of a stream of 16-bit digital *words*. These words are collected into *Frames* of 6 consecutive left/right pairs of digital samples. One frame therefore contains 192 audio bits which are then treated as a set of 24, 8-bit, *Audio Symbols*.

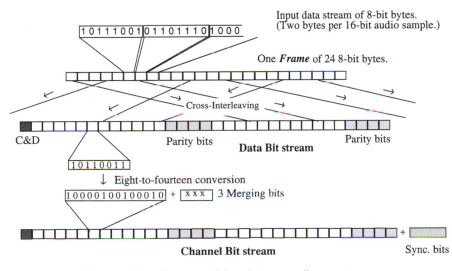

Figure 9.3 Compact Disc data encoding system.

These audio symbols are rearranged and some extra parity symbols are generated using an encoding scheme called a *Cross Interleaved Reed-Solomon Code* or CIRC. For our purposes it is sufficient to recognise that CIRC is a type of <u>block</u> code which generates a specific pattern of parity bits. CIRC also *Interleaves* or 'rearranges' the sequence of the data bits. The interleaving process is designed to minimise the effects of momentary data losses. Some extra *Control and Data* bits are also added at this stage. These contain extra information — for example track numbers and running time — which are of use to the CD player. The result of the CIRC encoding stage is to convert each frame from 24 audio symbols into 33 *data* symbols (each 8-bit, as before, giving a new total of 264 data bits). The parity bits provide some of the required ability to detect and correct random errors.

77

In practice, much of the data loss when replaying a CD occurs in brief *Bursts* when the player encounters a hole, or a piece of dirt, or when vibration causes the laser to momentarily miss tracking the data. This causes a series of successive data bits to be lost, sometimes lasting for a number of symbols. Interleaving or shuffling the symbols before recording (and de-interleaving them on replay) helps prevent successive audio symbols from being lost. It also 'spreads out' the data and parity bits to reduce the chance that both a given symbol and its associated parity bits will be lost. This interleaving process covers up to 28 <u>frames</u> and as a result, information from any pair of adjacent audio samples will usually be spaced some considerable distance apart on the actual CD. The usefulness of this interleaving process can be understood by considering the analogy of a piece of paper upon which a message has been typed. In the process of being passed to the person who wants to read it, the paper is attacked by a dog which tears it and eats a piece. As a result, when the message is read about 5% of the text is missing — perhaps because the last few lines have been torn off. It is likely that any information which was contained by the missing lines is lost (inside the dog!).

If the letters of the text had been typed onto the paper in a 'scrambled' order it would be possible to re-arrange the received text back into a message where occasional words would have a missing letter. (Of course, in order to do this the scrambling process must not be a random one as the person receiving the message has to know how to unscramble the text correctly.) The result would probably be a readable message despite the loss of letters from some words. This is because of the natural redundancy of the English language which lets us make sense of text even when there are mistakes. The CIRC encoding process works in a similar way. Parity bits are used to add some redundancy, and the message is interleaved (scrambled) so that any brief breaks in the data stream should only cause single-bit losses is some samples. These can then usually be corrected because of the redundancy.

Following CIRC encoding each of the 8-bit data symbols is translated into a 14-bit *Channel Symbol* and an extra three *Merging Bits* are tacked onto the end of these 14. For obvious reasons Philips refer to this process as *Eight to Fourteen Modulation* (EFM). At the end of the frame of channel symbols another 27 *Synchronisation Bits* are added to make a total of 588 channel bits per frame. The sync bits are a unique pattern which the CD player uses to locate the beginning of each data frame. The bits are then recorded as a sequence of *Pits* cut into the disc. Those parts of the disc surface where no pit has been formed are referred to as the *Land*. The recording is not made on the simple basis that '1'='pit' and '0'='land' (or vice versa). Instead a '1' represents the transition or <u>edge</u> between pit and land. A '0' means 'continue as before' and '1' means 'change from pit to land, or land to pit'.

The specific choice of which 14-bit channel symbol should represent each 8-bit data symbol has been made so as to try and satisfy a number of requirements. Firstly, a set of 14-bit codes has been selected whose patterns provide the largest possible *Minimum Hamming Distance* between adjacent codes. This helps the CD player recognise and correct occasional random bit-errors in the recovered data stream. There are 2^{14} = 16,384 possible choices of 14-bit channel symbols of which only 2^8 = 256 are required. We can therefore surround each legal pattern with 64 illegal ones.

Starting with an m-bit symbol, there are m ways of changing one bit to produce a new symbol. Any one of these new symbols could also be altered in m different ways by a second bit-change. However, this doesn't mean that we can produce m^2 <u>different</u> symbols by changing two bits since the second change will sometimes simply undo the first. Consider a typical initial 8-bit digital symbol, $abcdefgh$. For this example, m equals eight, so there are eight ways a one-bit change can produce a new symbol; $\tilde{a}bcdefgh, a\tilde{b}cdefgh, ab\tilde{c}defgh$, etc. (Here, the '~' above a character indicates that particular bit has been changed.) Symbols with two changes will be $\tilde{a}\tilde{b}cdefgh, \tilde{a}b\tilde{c}defgh, \tilde{a}bc\tilde{d}efgh,... a\tilde{b}\tilde{c}defgh, a\tilde{b}c\tilde{d}efgh,...$ etc. If we count up the numbers of symbols, $C\{m, q\}$, which differ from the one we started with by q bits, we find that

$$C\{m, q\} = \frac{m!}{(m - q)!\, q!} \qquad \text{...(9.8)}$$

As a result, if we allow <u>up to</u> Q bits of a symbol to change we can produce

$$N\{m, Q\} = \sum_{q=1}^{Q} \frac{m!}{(m - q)!\, q!} \qquad \text{...(9.9)}$$

new symbols which differ from the starting symbol by no more than Q bits. Now $N\{14,1\}$ = 14 and $N\{14,2\}$ = 105, hence, given that we can typically surround each legal 14-bit symbol with 64 illegal ones, we can expect to be able to use EFM to correct any single-bit errors and most double-bit ones.

The second factor which influenced the choice of 14-bit symbols was the decision to limit the maximum and minimum number of '0's which can

79

appear between successive '1's of the recorded bit stream. This sets a maximum and minimum distance between successive pit–land edges on the disc. (Remember that a '1' is recorded as a pit–land edge.) The codes chosen for CD recording ensure that there are always at least two '0's, and not more than ten, in between successive '1's. However, one symbol which finishes with a '1' may still need to be followed with another which begins with a '1'. The pair of symbols would then 'clash', violating the requirement for more than two zeros between any pair of '1's. This problem is overcome by the inclusion of three extra *merging bits* in between successive symbols. Now we can simply place three zeros in between symbols whenever we need to avoid a clash.

The symbols and merging bit patterns are also chosen to ensure that, on average, the encoded disc appears to the player as consisting of 50% land and 50% pit. This helps the servo control system in the CD player to correctly focus the laser spot it uses to read the recorded data.

Finally, the symbols and merging bits are chosen so as to produce a strong component in the recorded signal spectrum at a predetermined frequency. This provides a *clock reference* signal for the CD player. The player can compare a filtered version of the recovered signal with a crystal oscillator and use this to adjust the disc rotation velocity.

The encoding process converts an initial 192 bit frame into a recorded frame of 588 bits. The number of channel bits recorded on a 60 minute CD is, therefore, around 15·5 Gbits and the channel bit rate will be 4·32 Mb/s. This means that the actual channel bandwidth required must be over 2·16 MHz, <u>not</u> 0·7 MHz.

The ability of the CD system to withstand errors and disc or replay imperfections may be summarized in terms of four standard measures.

    i) *Maximum Completely Correctable Burst Length.* (MCL)

    = 4,000 data bits (2·5 mm of track length on disc.) This means that gaps or holes up to 2·5 mm across in an otherwise perfect disc should not lead to any loss of audio information. This indicates the power of the combination of the parity bits plus eight-to-fourteen modulation to correct the loss of a large number of successive channel bits.

    ii) *Maximum Interpolatable Burst Length.* (MIL)

    = 12,000 data bits (7·7 mm track length.) Once the MCL has been exceeded some data will become lost. The interleaving process is, however, designed to ensure that no two adjacent audio sample values will be lost until over 12,000 successive channel bits have become unreadable. The player can 'interpolate' the lost data samples.

The values for MCL and MIL quoted above assume that there are no other imperfections or random errors 'near' (i.e. within 28 frames) the error burst. A high random *Bit Error Rate* (BER) will degrade the above values. The effects of a given random bit error rate can be indicated by

Sample Interpolation Rate.

1 per 10 <u>hours</u> at a BER = 0·0001

1000 per minute at a BER = 0·001

This represents how often random bit errors conspire to overcome the error protection and make a sample value unrecoverable. When this happens the CD player can respond by interpolating the lost value from the adjacent samples. The rapid change in the interpolation rate with BER indicates a general property of digitised data communication. Given a reasonable degree of redundancy, a low level of random errors has almost no effect upon data reception. However, above some particular 'threshold' level the information loss rises dramatically with bit error rate.

Undetected Error Rate.

Less than 1 per 750 <u>hours</u> at a BER = 0·001

'Negligible' at a BER = 0·0001

This represents the frequency of undetected sample errors, i.e. the random noise produces a legal symbol and the required, equivalent, parity, which is not identical to that recorded. When this happens the CD player can't know that the recovered value is, in fact, wrong and an audible 'click' may result.

Summary

You should now know about the problem of *Aliasing* and how it can be prevented by using a filter <u>before</u> a signal is sampled. It should also be clear that — given enough bandwidth — a digital system can obtain a higher dynamic range than analog from a noisy channel. It should also be clear that the combination of a block-parity code (e.g. CIRC) and data *Interleaving* provides good protection against data loss due to random noise and burst errors due to missing channel data (soup on the CD!). You should also now know how the pattern of pits on a CD is calculated from the input signals.

Questions

1) Give an outline explanation of how a CD system encodes musical information into digital form and records it on a disc. Include an explanation of how the CD system protects information against random errors.

2) The CD system uses 16 bit samples and a *Sampling Rate* of 44,100 samples/second. What *Dynamic Range* and *Bandwidth* should this provide? How many bits of audio information will a 1 hour CD contain? **[96·3 dB. 5·08 GBits.]**

3) Explain what is meant by the term *Aliasing* and say what we must do to prevent it happening.

Chapter 10

The CD player as a measurement system

The CD player has to recover information from a spiral track of small pits which have been formed at a nominal *Information Layer* inside a compact disc. Unlike the old-fashioned vinyl (or shellac!) analog recordings, the CD does not have a 'continuous' groove and the optical sensor should never touch the disc. Hence the CD player must locate the required information without any mechanical guidance about where the data is to be found. Figure 10.1 illustrates the form of a typical CD surface and the optical beam used to read data from the disc.

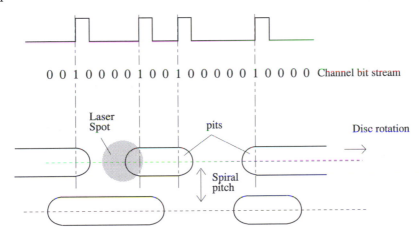

Figure 10.1   Replay of CD pit–land pattern.

Information is recorded on the surface of a CD in the form of a spiral track of pits, and is read using a laser whose wavelength is around $0{\cdot}7$ μm. The spiral pitch (distance between adjacent turns) is $1{\cdot}6$ μm and the disc is rotated so that the position illuminated by the laser spot moves at a constant <u>linear</u> velocity of $1{\cdot}25$ m/s.

Analog disc recordings were normally made at a constant <u>angular</u> velocity. This means that they can be replayed by rotating them at a steady rate. CDs use a constant linear recording velocity in order to maximise the amount of information which can be squeezed onto a given disc diameter. This means that the angular rate of rotation required to play a CD varies as the disc is played. Unlike most analog discs, CDs are recorded 'from the middle, outwards'. The optical sensor used to recover data starts near the middle of the CD with the CD being rotated relatively quickly. As the music plays the sensor moves outwards and the rotational rate is reduced.

CDs can be manufactured in various ways. Many of the first discs were made using photochemical techniques. A light sensitive chemical was coated onto the surface of a disc of plastic. The required pit–land pattern was then 'photographed' onto the disc. The details of this pattern were then etched using appropriate chemicals. More recently, faster, cheaper methods have been developed. For example, many modern discs are produced by *Injection Moulding* — forcing plastic into a metal mould, one wall of which holds a 'negative' version of the required pattern of pits and land. The patterned plastic surface is coated with a thin layer of metal (usually aluminium, but some expensive CDs use gold instead) to make it highly reflective. This is then covered with a protective top coating of transparent plastic.

When using electromagnetic radiation to observe small-scale features, we wouldn't normally expect to be able to measure anything whose size is significantly smaller than the chosen wavelength. In the case of a CD the required bit recovery rate is 4·32 Mbits/s and the the disc velocity is 1·25 m/s. This implies that each bit occupies a track length of just 0·29 μm — i.e. less than half the laser's wavelength! A number of factors help CD players to recover information from such a closely packed surface pattern.

- Firstly, the laser beam is tightly focused to produce a spot whose nominal diameter is typically around 1 μm. This requires an optical system of very high quality.
- Secondly, the encoding system is designed to help the laser sense the surface features. Every stretch of pit or land will be at least 3 bits long. This is a result of the coding requirements that; i) there must always be at least 2 zeros between adjacent ones; ii) pit–land edges represent encoded 1's. This means that pit–land edges will always be at least 0·87 μm apart — i.e. the length of each pit or land feature will always be comparable with the laser wavelength. This means it is possible to ensure that the laser spot will never illuminate more than a single edge at a time.
- Thirdly, the optical system employs a highly coherent light source and the pits are made approximately a quarter-wavelength deep. The readout beam axis is nominally aligned to be perpendicular to the disc plane. When there are no pit–land edges in the spot, all of the

84

reflected beam will share the same phase. The phase of the reflected beam will, however, change by 180 degrees when the spot moves from pit to land, or vice versa.

When the optical spot traverses a pit–land edge the magnitude of the beam reflected back into the sensor optics will momentarily dip almost to zero. The reason for this can be understood by considering what happens if half the spot energy falls upon land, and half into a pit. The reflected beam then consists of two portions, equal in magnitude but opposite in phase. As a consequence the total energy coupled back into the sensor beam would be zero. Of course, the 'missing' energy does not just vanish, instead it is scattered in some other direction, away from the sensor beam.

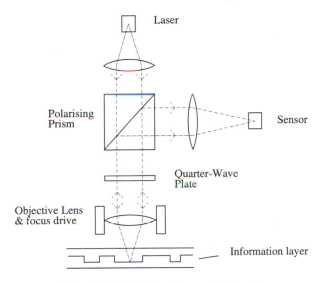

Figure 10.2 Typical CD replay optical system.

Although the details vary a great deal from one manufacturer to another, most players use variations on the system illustrated in figure 10.2. In principle it would be better to employ some form of *Michelson Interferometer* with a pair of detectors. This would enable the player to measure the phase of the reflected signal as well as its amplitude and distinguish pits from land. This would improve the S/N ratio achievable with a given laser power level. However — as will become clear later — the player's optical system is invariably much more complex than implied by figure 10.2. The extra complexity is to allow for the chosen focusing/tracking arrangements. The use of a full interferometer system would require a further increase in player complexity. Fortunately, poor signal to noise ratio need not be a problem with CD players as the manufacturer can generally obtain solid state laser sources which provide ample power levels. Hence the use of a full phase interferometer system isn't usually regarded as necessary. (Although doubtless some manufacturer will eventually make it's inclusion a 'selling point'!)

The system illustrated relies upon detecting the momentary dips in the observed reflected light level which occur at the pit–land edges. Laser light is focused onto the disc information layer via a polarisation prism and a quarter-wave plate. Since this isn't a book on optics we don't have to get into an explanation of just how these items work. For our purposes it's enough to know that a polarisation prism will transmit light with one plane polarisation and reflect light polarised at 90 degrees to the transmission plane. The quarter-wave plate alters the polarisation state of light passing through it. As a result, the light reflected back from the disc is directed onto a sensor, not returned to the laser.

In practice we may find that the reflected energy is not divided exactly 50:50 at the pit edges. The pit depths may also not be exactly a quarter wavelength. This means that the magnitude of the sensed reflection may not dip right down to zero. Despite this practical problem, the power of the replay laser is normally so large that we can obtain a high enough S/N ratio to determine the locations of pit–land edges with an uncertainty considerably smaller than a wavelength.

When the system is working correctly, the laser spot is focused on the information layer which sits in the nominal information layer of the disc surface. (This layer can be defined to be mid-way between the land and pit bottom planes.) Light reflected by the disc will return through the system and be refocused at the required output plane, just in front of the signal detector (or detectors).

Any fluctuations in the distance from the objective lens to the information layer will have two undesirable effects. The beam size at the information plane will become larger, and the output focal spot will shift along the beam axis away from its required position. The CD player must, therefore, be able to continuously adjust the objective lens position to maintain its position at the correct distance from the disc. It must also ensure that the spot tracks the spiral pattern of pits.

Since there is no physical contact, the CD optical sensor system must itself provide signals which can be used to continuously adjust its position relative to the disc with sub-micron accuracy — even when playing a disc which is warped or rotating off centre by over a hundred times this amount. It must also provide a measurement of tracking velocity with enough accuracy to enable the player to vary the disc rotation rate and collect audio data with a channel bandwidth of over 2 MHz. The CD player must therefore contain a highly accurate and responsive position/velocity measurement system.

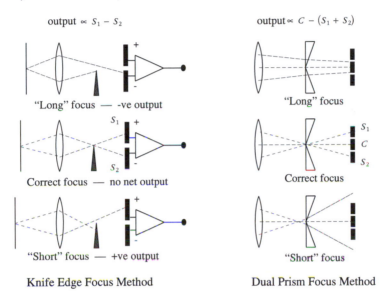

Figure 10.3 CD spot focus mechanisms.

CD player manufacturers have used a variety of techniques to control the optical recovery of information and position the sensor correctly. For the sake of illustration we can examine four techniques:

i) dual-prism focusing	ii) knife-edge focusing
iii) three-spot tracking	iv) dither tracking

Figure 10.3 shows the focusing methods we are considering. In each case the output light level is detected by using more than one sensor. The systems are arranged so that any alteration in objective-disc spacing alters the relative levels seen at the sensors. In the *Knife-Edge* system, an opaque edge is placed near the output focal spot. This stops some of the light from reaching a pair of sensors placed a little further along the beam.

When focused correctly, the output focal spot rests near to the knife-edge and the amounts of light reaching each sensor of the pair of output sensors, S_1 and S_2, are reduced by similar amounts. Any change in the objective-disc spacing will shift the output focal plane along the beam, producing an imbalance in the amounts of light blockage experienced by each sensor.

The output voltages produced by the pair of sensors can be monitored by the CD player. The sum of their voltages, $S_1 + S_2$, can be used to provide audio information. Any <u>difference</u>, $S_1 - S_2$, in their voltages can be used to indicate a focusing error. The sign of this difference voltage indicates the direction of the error. When this difference output is zero the system is ideally focused

The *Dual-Prism* system employs a pair of prisms placed in front of three light sensors. The prisms slightly alter the convergence of the beam, changing the relative levels falling upon <u>three</u> output sensors. The size of the effect of the prisms depends upon the position of the focal plane of the incident beam relative to the prism. As with the knife-edge system, we can use the sum of all the sensor voltages, $C + S_1 + S_2$, to obtain audio information. The difference, $C - (S_1 + S_2)$, between the central sensor and the surrounding ones can be used to indicate any focusing error. As with the knife-edge system a zero difference output indicates when the system is ideally focused. When this happens we can expect about half the light power to be falling on the central detector, C, and the other half on the outer pair. This means that about a quarter of the total falls on <u>each</u> of S_1 and S_2. A focus error in one direction will cause C to rise and $S_1 + S_2$ to fall. An error in the other direction has the opposite effect. Hence, as with the knife-edge system, the sign of the difference output indicates the direction of the error.

However the focusing information is gathered it provides the player with a focus control signal whose magnitude and sign depend upon the amount (and direction) by which the objective-disc spacing differs from the required value. This signal is then amplified and used to drive a motor which changes the objective lens position so as to reduce the error. The overall system acts as a form of *Servo Control Loop* to maintain the required focus.

Figure 10.4 shows the *Three–Spot* method for obtaining tracking measurements. In this system the laser beam is diffracted so as to produce three spots focused on the information layer of the disc. The power reflected at each spot is directed onto a separate light sensor. The spots are arranged to lie in a line at a slight angle to the nominal direction of the information spiral. As a result, when the centre spot is correctly aligned the front and back spots only partly illuminate the spiral.

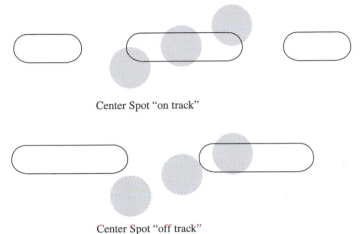

Center Spot "on track"

Center Spot "off track"

Figure 10.4   Three–spot spiral tracking system.

The CD player monitors the relative levels of light modulation recovered by all three of the spot sensors. When the system is tracking ideally, the centre spot will give a relatively large modulation output. When the spots are slightly off-track the output from either the front or back spot will increase and that from the centre one will fall. The difference in levels between the front and back spot sensors can therefore be used to obtain a measure of any tracking error. As with focusing, any difference signal can be amplified and used to adjust the position of the objective lens so as to maintain good tracking.

A disadvantage of this system is that only about one third of the available laser power will be used to obtain the required audio information. In principle, the player could recover audio information from all three spot sensors. A problem with attempting this is that the spots are looking at different places along the spiral track, and hence at any moment they are recovering different portions of the recorded data. The spots could, if we wished, be placed 'side by side' to overcome this problem, but they would then be physically overlapping and — as a result — the sensors would be more likely to see light coming from the 'wrong' spots.

The *Dither Tracking* technique makes a single spot do the work of three by forcing the spot to hunt back and forth across the spiral track. This can be achieved by vibrating the objective lens from side to side a very small amount, or by reflecting the laser-sensor beam off a mirror surface whose angle is vibrated. Typical systems employ a sinusoidal modulation with a frequency of a few hundred Hertz. The magnitude of the oscillation is very small and should only move the spot at the disc information layer by a fraction of a micron.

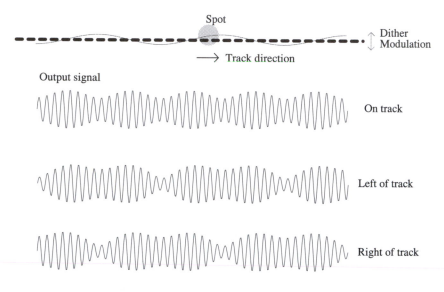

Figure 10.5   Using spot 'dither' to obtain tracking information.

The effect of this deliberate modulation is shown in figure 10.5. The output signal mainly consists of a complex signal with frequency components in the MHz range. When the system is tracking correctly, the dither produces a slight amplitude modulation of this signal at twice the dither frequency. Any tracking error will change the shape of this modulation, producing a shape which contains a component at the dither frequency.

In an earlier chapter we saw that the digital signal recorded on the CD requires a channel frequency range up to at least 2·15 MHz to cope with the required bit-rate. One of the conditions used to select the coding system was that there should never be more than ten '0's between any successive pair of '1's in the data stream. This means that the frequency spectrum of the digital signal won't contain any significant frequency

90

components below about $2 \cdot 15/10$ MHz ≈ 200 kHz. The digital audio data is therefore confined to a frequency band from about 200 kHz to 2 MHz. Since the dither frequency (and twice this frequency) is well below the digital audio frequency range the CD player has no problem separating the dither tracking output from the digital audio information. The player can then compare the magnitudes of the amplitude variations at the dither and $2 \times$ dither frequency, find the phase of these signals relative to the dither modulation it is applying, and use the result as a measure of any tracking error.

The dither technique is, in principle, a very efficient one. Only one spot and sensor are required, and the magnitude and frequency of dither can, if we wish, be continuously altered to suit the difficulty of the task (i.e. less dither on 'good' discs and players). Perhaps the main drawback of this method is that its name makes it easily confused with the (quite different!) dither 'noise' signal used to suppress quantisation distortion. Various other methods have been devised by manufacturers to recover tracking information. However, as with the choice of focusing technique, what finally matters is the quality of the actual CD player design and manufacture.

Summary

You should now understand how the pattern on the surface of a CD is formed. You should also know how a CD player is able to 'track' and recover the spiral of pits in the CDs information layer. In particular, it should be clear how the player can focus and align its laser/sensor system and adjust the rotation rate to recover the required stream of channel bits.

Chapter 11

Oversampling, noise shaping, and digital filtering

11.1 The CD player as a digital signal processing system

The stream of bits recovered from the disc is processed through a series of stages which reverse the encoding process which occurred when the signals were recorded. Minor errors can be completely corrected using the eight-to-fourteen redundancy and parity checking built into the system. Major errors may result in the unavoidable loss of information, but most CD players then use a pre-programmed algorithm to 'fill in' or *interpolate* occasional lost samples. The details of this algorithm for masking information loss will differ from one player to another. The recovered stream of digital values can then be passed to two digital to analog convertors (DACs) for conversion into an output pair of analog audio signals.

In principle, we could simply use the CD player to recover 44,100 pairs of digital samples per second and employ a pair of 16-bit DACs to obtain analog signals. Whilst this approach would have the advantage of simplicity it may produce an output which exhibits the 'staircase' distortions mentioned earlier.

Provided the input signal was dithered before sampling, any staircase distortions can — in theory — be removed by passing the output from the digital to analog convertors through low-pass filters which reject frequencies above half the sampling frequency. This is because, in an ideal system, all the unwanted frequencies produced by the staircase effect will be above 22·05 kHz. Some of the earliest CD players did employ this approach, but it soon proved unsatisfactory for a variety of reasons and has largely been superseded by better methods. Generally speaking, simply using analog filters to 'clean up' the output waveforms works poorly for two reasons:

Firstly, the CD player (or the information on the disc!) may be imperfect. For example, any production problems in manufacturing the digital to analog convertors will alter the form of the staircase distortion and may

92

produce unwanted components inside the analog signal's frequency range.

Secondly, in order to realise the full potential of the CD encoding system we would require low-pass filters which perform amazingly well. Ideally, they should pass any signal frequencies up to almost 22·05 kHz without altering them in any way, but must reject any distortion components above 22·05 kHz by <u>at least</u> 95 dB to prevent them from degrading the potential dynamic range. Analog filters capable of simultaneously meeting both these requirements can be made. However, they are difficult to produce as they must contain a large number of very accurately toleranced components. This makes them large and expensive. It is also inevitable that the values of some components will tend to change with age, temperature, or humidity. This would mean a very expensive CD player whose performance might deteriorate audibly with use.

To avoid these problems, almost all modern CD players process the digital data in some way before presenting it to the convertors. The main objects of this processing are:

- To perform a computation equivalent to low-pass filtering. This is intended to reduce the severity of the staircase distortions, easing the demands imposed upon any analog filters placed after the convertors.
- To help prevent any imperfections in the digital circuits, especially the digital to analog convertors, from producing other signal distortions.

The details of this digital processing vary considerably from one type of player to another. (And, of course, every manufacturer claims to use the 'best' method for their newest models!) Fortunately, all of these processes are aimed at achieving the same end result so we need only consider one example. Here we will look at the original system employed by Philips in their 'first generation' CD players using the SAA7030 and TDA1540 integrated circuits. The following explanation has been simplified to some extent, to make it easier to follow, but contains the essential features of the process.

This system employed a combination of two techniques, *Oversampling*, and *Noise Shaping* to achieve the desired results. Oversampling means that a set of sampled values is used to calculate the values we 'would have obtained' at intermediate moments if the original input had actually been sampled more frequently. Provided the sample values we start with satisfy the

93

sampling theorem these extra values don't contain any new information. This is because there is only one possible waveshape which can fit the sampled values read from the CD. The first Philips CD players employed ×4 oversampling, converting an input data stream of 44,100 samples/ second (per channel) into 176,400 samples/second. We can regard staircase distortion as being an unwanted high-frequency variation which has been added onto the signal we wish to communicate via CD. By ×4 oversampling we produce the effect shown in figure 11.1.

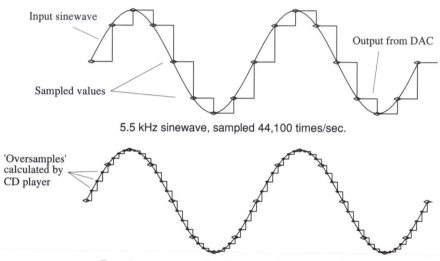

5.5 kHz sinewave, sampled 44,100 times/sec.

Four times oversampling the reconstructed waveform.

Figure 11.1   Effect of × 4 oversampling.

For the sake of the illustration we can consider an input signal in the form of a 5·5 kHz sinewave which is sampled by the CD recording process at 44,100 samples/second. The simplest way to convert these sampled values back into an analog waveform would be to use a digital to analog convertor (DAC) which produces an output level appropriate for each sample and then 'holds' this level until it is time to output the next sampled level. This kind of output is called *Sample and Hold* and produces the kind of staircase distortion shown.

The sampling theorem says that, provided a series of samples form a complete record of the original information, we can use the measured sample values to calculate the actual signal level at <u>any</u> moment in between the sampled instants. These calculated values can then be given to the DAC in between the 'genuine' samples to produce the improvement showed in the lower waveform of figure 11.1. It is important to realise that these calculated samples are <u>not</u> 'guesses', but really do represent the signal level which <u>would have been observed</u> if the input waveform had been measured at these moments. If the CD recording process had actually recorded these extra values on the disc the player would not be provided with any extra information since the original set already contain a complete record of the waveform information. Hence the term '*oversamples*', which indicates that these extra values — calculated or measured — don't actually contain any fresh information.

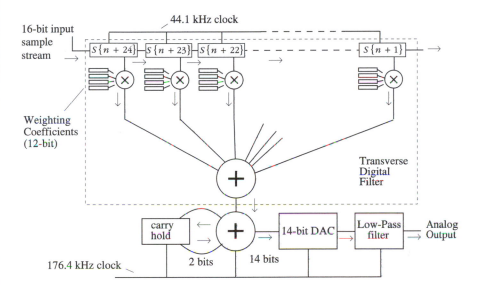

Figure 11.2   Schematic diagram of Philips SAA7030 + TDA1540.

The use of × 4 oversampled digital values reduces the staircase effect in two ways. The basic frequency of the unwanted staircase distortion is increased by a factor of 4, and its amplitude is reduced by a factor of 4. As a consequence it becomes much easier to produce analog filters which, placed after the DACs, will suppress this distortion without significantly affecting the wanted signal. Figure 11.2 shows a schematic diagram of (one stereo channel of) the initial Philips processing system. This used two integrated circuits (ICs), the SAA7030 and TDA1540. The 16-bit samples read from the disc are clocked through a serial *Shift Register* which, in this case, can hold 24 successive sample values. The rate at which the system processes the data samples is determined by two *Clock Frequencies*, 44·1 kHz and 176·4 kHz, which are supplied to the ICs. These two clock signals are *Phase Locked* so that every fourth cycle of the 176·4 kHz starts at the beginning of each 44·1 kHz cycle. In figure 11.2, $S\{n + 24\}$ represents the 'newest' sample value and $S\{n + 1\}$ represents the 'oldest'. (It's assumed that n samples have already passed through the system and have been discarded.)

The 44·1 kHz clock signal is also used to control the rate at which digital samples are recovered from the CD, hence samples should be presented to the input end of the shift register at the same rate they are read from the disc. At the beginning of each 44·1 kHz clock cycle all the sample values stored in the register locations are shifted along one place. A new sample is entered into the first register location and the 'oldest' sample value is thrown away from the last register location. The registers are linked to an array of multiplier circuits. Each of these has a set of four coefficient values connected to it. These coefficient values are usually built into the processing IC when it is manufactured, although some modern CD systems allow the coefficients to be modified by replacing or reprogramming a ROM (memory) chip.

The 176·4 kHz clock controls the data processing carried out by the circuit. Each *Processing Cycle* takes one 44·1 kHz clock period — i.e. four 176·4 kHz clock periods. To see how the system operates we can examine what happens during each of the four 176·4 kHz clock periods of a processing cycle.

At the start of the first period all the samples are shifted along and a fresh sample value is entered at the 'newest' end of the line of registers. This event triggers the start of the processing cycle. Immediately after the sampled values have been updated they are all multiplied by the first set of coefficients and the results are added together to produce an output value which is sent forward to the digital/analog convertor system.

During the second 176·4 kHz clock period the register values are multiplied by the second set of coefficients, and the results added, to produce a new output value for the convertor. The third set of coefficients are used during the third 176·4 kHz clock period, and the last set during the fourth and final period of the processing cycle. As a result, each processing cycle produces four distinct output values which are sent to the convertor. These have all been obtained from the same 24 input samples, but used four distinct sets of coefficient values. During the next processing cycle the input data is shifted along, a new sample is injected, and the process is repeated to produce four more output values.

96

In Chapter 7 we saw how it is possible to recover the signal value at instants in between samples. Here the action of the IC may be seen as carrying out a similar task. We could therefore use a set of coefficients which correspond to the values of the sinc function indicated by the expressions in Chapter 7. This would serve to 'smooth out' some of the output signal distortion effects. In practice, however, it can be an advantage to slightly alter the coefficient values to obtain a flatter frequency response, lower distortion, or whatever we require.

The circuit which carries out this process (an SAA7030 in this case) is called a *Transverse Digital Filter* (TDF). By choosing an appropriate set of $4 \times 24 = 96$ coefficients we can carry out a series of computations which mimics the effect of a '96'th order' analog filter. The frequency response of this filter depends upon the values chosen for the coefficients. In theory we could build an analog filter, using capacitors, inductors, etc, to achieve the same end. This is because, in principle, identical results can be achieved by either analog or digital processing of the information. In reality, of course, the analog equivalent would prove far more difficult to make, and its properties would be relatively unstable. There are, therefore, good practical reasons for carrying out this filtering process in the digital domain.

One interesting consequence of using a set of samples to compute output values is that the results have more bits per value than the input samples! The SAA7030 stores its internal coefficients as 12-bit numbers and the output values obtained from the TDF therefore emerge as $16 + 12 = 28$-bit numbers. Note that these additional bits don't contain any 'new' information. They are a consequence of the way information is 'redistributed' by the TDF process. In effect, each bit of real data influences more than one bit of the oversampled results. The oversampled bits aren't all 'independent' of one another.

In an ideal world we might choose to employ a pair of 28-bit DACs after the filter. Alas, at the time the first CD players were launched Philips were doubtful that they could mass produce even 16-bit convertors of the required precision at a commercial price! They could, however, make good 14-bit convertors able to run at a clock speed of 176·4 kHz. They therefore decided to use 14-bit DACs in the first generation of CD players. At first glance it seems that the use of a 14-bit convertor will unavoidably cause some audio information to be lost. Fortunately, it is possible to process the data before conversion in a way which can prevent any information loss by using a *noise shaping* technique. As with oversampling this process may be carried out in various ways. The original Philips design employed a method which we can understand by looking at figure 11.2.

The 28-bit output values from the TDF are passed through an adder into another register. The most significant 14-bits are then sent on to the DAC. The unused, least significant, bits are treated as a 'remainder' which is held for one 176·4 kHz clock period and then returned to the adder to be combined with the <u>next</u> value. This process is repeated with each successive value. This has the effect of 'carrying forward' any error between the converted and presented values. To see how this works we can forget, for a while, the potential ability of the TDF to generate 28-bit numbers and consider what would happen if we just present a series of 16-bit values to the noise shaping system. For simplicity, imagine that four successive values are the same and let F represent the most significant 13 bits of their value. (We'll also assume that we start with a carry of zero from the last cycle.) The 'carry forward' process continued over four clock cycles then looks like:

16-bit input	+Carry	14-bits to DAC	Remainder
			00
F001	F001	F0	01
F001	F010	F0	10
F001	F011	F0	11
F001	F100	F1	00

Note that if we add together the four successive 14-bit output values sent to the DAC we obtain F001 once again. A low-pass filter placed after the digital to analog convertor will have the effect of suppressing any short-term fluctuations in the output level. If this filter attenuates frequencies above half the basic sampling rate (i.e. 1/8th the oversampled rate) it will tend to produce an output which is much the same as if we had averaged the four values, producing an output equivalent to that which would have been produced by a 16-bit convertor.

It is perhaps unfortunate that this process has come to be called noise shaping as the name implies that the process is somehow 'random'. In reality the process operates by attempting to average away the *Truncation* effects produced by the finite number of bits per digital value. It does this by storing any truncation errors and using them to adjust later output to produce a more accurate overall output.

For the sake of the above explanation we ignored the fact that, using 12-bit coefficients, the TDF is capable of providing 28-bit output values. Some manufacturers of CD players have taken advantage of this by employing DACs which convert 18, 20, or even more bits per sample in an attempt to produce more 'accurate' analog output signals. It is important to realise that, although this process can provide a 'smoother' output waveform it doesn't magically produce any extra information which wasn't in the original set of 16-bit samples. In principle, an 'ideal' 16-bit DAC and analog filter would produce the same results as any other 'ideal' noise shaped and oversampled system. Any differences stem from how well the system is designed and built, not from any inherent theoretical differences.

Summary

You should now know what is meant by the terms *Oversampling* and *Noise Shaping*. That these are digital signal processing techniques which can be used to perform functions similar to filtering an analog signal. You should also now understand how a *Transverse Digital Filter* works. It should also be clear that — in theory — the same results can be achieved using systems which produce anything from one to umpteen bits per value presented to the output DACs <u>provided</u> that the digital process is performed correctly.

Chapter 12

Analog or digital?

12.1 Is the world 'analog'?

In general, we can imagine representing information in terms of some form of analog or digital signal. The digital data stored on a CD will normally have been produced using analog to digital convertors which are fed with amplified signals from microphones. The original microphone signals are obviously 'analog' — or are they?...

Modern physics is largely based upon the concept that the world behaves according to the rules of *Quantum Mechanics*. One of the axioms of this is that <u>all</u> forms of energy behave as if quantised. This gives us the well-known (although not well understood!) 'wave–particle duality'. Statistically, the behaviour of physical processes can be described in terms of things like waves and continuous functions. Yet, when we examine any process in enough detail we can expect to see behaviour which it is more convenient to describe in terms of distinct particles or 'packets' of energy, mass, etc.

When the Compact Disc system was originally launched some people criticised it on the grounds that, 'Sound signals are inherently analog, i.e. sound is a smoothly varying (continuous) pattern of pressure changes. Converting sound information into digital form "chops it up", ruining it forever.' This view is based on the idea that — by its very nature — sound is inherently a wave phenomenon. These waves satisfy a set of *Wave Equations*. Hence we should always be able to represent a given soundfield by a suitable algebraic function whose value varies smoothly from place to place and from moment to moment. Since the voltage/current patterns emerging from our microphones vary in proportion to the sound pressure variations falling upon them it seems fairly natural to think of the sound waves themselves as having all the properties we associate with 'analog' signals, i.e. the sound itself is essentially an analog signal, carrying information from the sound sources to the microphones. But how <u>can</u> sound be 'analog' if the theories of quantum mechanics are correct?

The purpose of this chapter is to show that the real world isn't actually either 'analog' <u>or</u> 'digital'. Analog and digital signals are no more than

100

mathematical representations of reality, useful when we want to process information. In fact we could say the same thing about the 'waves' and 'particles' we use so much in physics. Although it's easy to forget the fact, both waves and particles are mental models or 'pictures' we use to help us grasp how the real world behaves. Although useful as concepts, they don't necessarily 'really exist'. To illustrate this point, imagine a situation where we are given a working electronic circuit board without being told anything about it and asked, 'Is this an analog or a digital circuit?' How could we tell? Of course, we could probably decide by looking to see if the circuit contained any integrated circuits, reading their type numbers, and looking them up in a book! (We can also guess that if the circuit doesn't contain any integrated circuits, it's probably not digital...) However for our purposes, this would be cheating. The real question is, 'Can we tell just by looking at the kinds of electronic <u>signals</u> being passed around between components on the board?'

If we connect an oscilloscope we can watch how some of the voltage or current levels in the circuit vary with time. In most cases, the shapes of the waveforms we'd see on the oscilloscope would quickly show whether the signal was digital or analog.

Digital signals will often show 'square' shapes. The signal voltages tend to spend most of the time near one or the other of two particular levels, switching between them relatively quickly. Analog signals sometimes show no obvious patterns, although in some cases they show a simple recognisable shape like a sinewave. As a result we can sometimes form an opinion about the type of signal by seeing if we can recognise the waveforms. But is there a more 'scientific' — i.e. objective — way of deciding? Is their an algorithm or recipe which would always be able to tell us what form a signal is taking?

At first it might seem as if this problem is an easy one. When we look at them on an oscilloscope, digital signals can look nice and square, analog ones tend to look like bunches of sinewaves or noise. Unfortunately, when an information channel is being used to its limits the situation can be less clear. When a digital signal is transmitted at very high bit-rates, the rising and falling edges of each level change tend to become rounded by the finite channel bandwidth. As a result, the actual transmitted voltage fluctuations may not display an obviously digital pattern.

In a similar way, some analog waveforms may show fairly square patterns. For example, the output from a heavy rock band, compressed by studio

101

equipment, can have a 'clipped' look similar to a stream of, slightly rounded, digital bits. Also, if an analog channel is being used *efficiently* every possible waveform shape will appear sometimes. As a result, the waveform will sometimes look just like a digital one.

We can't know with absolute certainty, just by examining a real signal pattern for a while, whether it carries information in either digital or analog form — although we can be fairly confident in many cases. We use voltage patterns (or currents, etc) to carry information in various ways, but the terms 'digital' or 'analog' really refer to the way we <u>process</u> information, not some inherent property of the voltage/current itself.

For most purposes this lack of absolute knowledge doesn't matter. But it serves to make the point that digital and analog signals are <u>idealisations</u>. Any real signal will have both analog and digital characteristics.

12.2 The 'digital' defects of the long-playing record

In the previous section we considered the signals used to communicate information. But what about the physical processes and sensors we use to create or collect information? In general we tend to assume that a measurement system operates in an analog manner. An input is sensed by some form of detector and produces a voltage or current whose magnitude varies in proportion with the stimulus. This voltage or current is then taken as an *analog* of the input we wish to measure.

Despite this assumption we can expect that any physical process must, at some level, be affected by the quantum mechanical behaviour of the real world. In order to see how this influences a real measurement we can consider the example of a Long Playing (LP) record. This sound recording system makes a useful contrast to the Compact Disc which we have already examined. It is also considered by some Hi-Fi audio enthusiasts opposed to digital audio as a paragon of 'analog virtues'.

Information is stored on an LP in the form of a modulated spiral groove pressed into its surface. The measurement sensor consists of a stylus which is placed in the groove whilst the LP is rotated at a constant angular velocity. An output signal is produced which is proportional to the instantaneous radial velocity of the stylus The signal is recorded in the shape of the groove surfaces, or 'walls'. The stylus is connected to some form of electrical generator (usually a coil in the vicinity of a magnet)

102

which produces an output voltage proportional to the transverse velocity of the stylus. In general, sensors which convert one form of energy into another are called *Transducers*. In this case some of the rotational energy of the LP is converted into electronic energy. The combination of stylus and generator is usually referred to as a 'cartridge'. (It can also be called a 'pick-up', but this term is confusing as it's sometimes used for the arm which supports the cartridge above the LP record.)

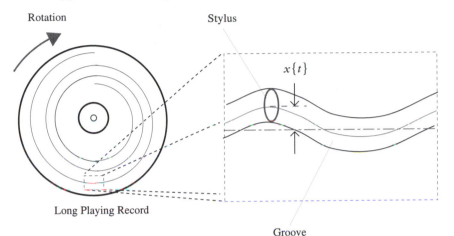

Rotation   Stylus

$x\{t\}$

Long Playing Record

Groove

Figure 12.1   Conventional view of LP groove and stylus.

For the sake of simplicity we can assume that the LP is *Monophonic* and that the nominal centre line of an unmodulated groove would cause the stylus to move inwards at a constant rate, $\frac{dr}{dt}$. We can represent the recorded signal as illustrated in figure 12.1 by an offset distance, $x\{t\}$, between the actual position of the stylus at time, t, and the position it would have if there were no modulation. The radial velocity of the stylus, $v\{t\}$, of the stylus at any instant will be

$$v\{t\} = \frac{dx\{t\}}{dt} + \frac{dr}{dt} \qquad \qquad ...(12.1)$$

In practice the steady spiral velocity, $\frac{dr}{dt}$, simply causes the pick-up arm to move slowly inwards so we can say that the output voltage generated by the stylus movements will be

$$v\{t\} = kv\{t\} = k\frac{dx\{t\}}{dt} \qquad \qquad ...(12.2)$$

where k is the appropriate conversion coefficient (the cartridge's *Sensitivity* or *responsivity*) of the cartridge. For a real LP system, k is typically in the range 0·1 – 1 mV/cm/s. Ideally, we would like to obtain an output signal,

103

$v\{t\}$, which is a faithful reproduction of the required sound pressure variations. In any real system, however, some problems must be taken into account. For example, various processes will restrict the dynamic range of the system. Mechanical problems will place limits on the maximum possible size of the displacement, $x\{t\}$, and the maximum achievable acceleration, $\frac{d^2x\{t\}}{dt^2}$. The noise level will also prevent us from observing changes in displacement smaller than a given size.

The record industry adopted a standard level of 5 cm/s (peak velocity for a 1 kHz sinewave), as the nominal 0 dB *Reference Level*. A reasonably good cartridge would have been able to *Track* (maintain its stylus in the groove) modulation levels around 20 dB greater than this reference level. For a sinewave of frequency, f, amplitude, A, the offset displacement will have the form

$$x\{t\} = A \, \mathrm{Sin}\{2\pi f t\} \qquad \qquad \dots (12.3)$$

hence the velocity will be

$$v\{t\} = 2\pi f A \, \mathrm{Cos}\{2\pi f t\} \qquad \qquad \dots (12.4)$$

and the acceleration

$$a\{t\} = -(2\pi f)^2 A \, \mathrm{Sin}\{2\pi f t\} \qquad \qquad \dots (12.5)$$

A 1 kHz sine wave recorded at a +20 dB level will have a displacement of peak value, $x_{peak} \approx 80 \ \mu\mathrm{m}$, and a peak acceleration, $a_{peak} \approx 3$ km/s/s. (i.e. a peak acceleration around 320 times bigger than that due to the Earth's gravity!)

No matter how well they have been made, every cartridge will 'mistrack' groove modulations above a given magnitude. This is usually because the accelerations and displacements become too large and the stylus either loses contact with the groove walls or gouges into them, damaging the record! In other cases the stylus may remain in contact, but the cartridge's electrical output saturates. Whatever the exact cause, above a given level the cartridge (sensor) output ceases to be a faithful representation of the groove modulation. These electro-mechanical problems will limit both the maximum signal level and the maximum rate of change of the signal level we can obtain using a given cartridge.

The smallest signal levels we can sense using the cartridge will be partly set by electronic noise produced in its generator resistance and in the amplifier used to boost its output. There is also a <u>mechanical</u> limit on the smallest signal level which will be clearly measurable.

104

A 0 dB 1 kHz sinewave corresponds to a peak offset, x_{peak}, of just 8 µm. An LP record is made from a solid assembly of real atoms and molecules. In practice, LPs are made of an amorphous polymer, *PolyVinyl Chloride* (PVC), to which various other materials have been added. The precise properties of this material are quite complex and were the subject of quite a lot of research and development by the music industry (tobacco-ash, insects, etc, have also been found in LP material!). To avoid the complexity of the details of PVC's properties we can imagine an LP made of crystalline carbon (diamond!). It must be admitted that manufacturing such an LP would be rather difficult!

The walls of the groove of such an LP would be made from layers of carbon atoms. Each carbon atom has an effective diameter of around half a nanometre so the thickness of each layer will be approximately 0·5 nm. The position of the stylus is determined by resting on top of the uppermost layers of atoms. Hence we can see that the stylus position will be roughly *quantised* by the finite thickness of the atomic layers. When playing a sinewave whose peak size is 8 µm the movement of the stylus would take place in 1 nm steps. Instead of smoothly varying, the stylus offset would therefore always adopt one of the set of available levels, $x\{t\} = m.\Delta x$, where m is an integer and Δx is the thickness of the atomic layers. The effect is to divide the ±8 µm swing of a 0 dB 1 kHz sinewave into 32,000 steps — just as if the signal had passed through an ADC!

If we assume that the largest possible recorded signal level is +20 dB (i.e. $x_{peak} = 80$ µm) and accept that the signal is quantised in 0·5 nm steps then the diamond LP has a dynamic range, D, of

$$D = 20. \, \mathrm{Log}_{10}\left\{\frac{2x_{peak}}{\Delta x}\right\} \approx 110 \text{ dB} \qquad \qquad ...(12.6)$$

This compares very well with the Compact Disc system which employs 16-bit digital samples and hence has a dynamic range of about 96 dB. Alas, the performance of a real LP and stylus may be very different from the imaginary example! The actual dynamic range of a real LP is normally much less than 100 dB!

PVC is a *Polymer*. This means its molecules have been grown by joining together lots of smaller molecules. The results of this polymerization process will depend upon the details of the process. The average molecular weights of the polymer chains which are formed can range from a few tens of hydrogen atom masses to hundreds of thousands. As a result, the PVC molecules are much larger than carbon atoms. This has

105

the effect of producing a material which is 'lumpy' with a typical quantisation size far bigger than a carbon atom. As a result, the value for Δx we should have used for the above expressions is hundreds of times larger than 0.5 nm, producing a much smaller dynamic range.

The purpose of the above example was to help us recognise that, since LPs are made from a collection of real molecules, the signals they hold <u>must</u> be quantised. Fortunately for the LP this usually isn't obvious. The underlying signal quantisation is usually masked by various effects.

Although the PVC molecules are much larger than carbon atoms they <u>aren't</u> arranged into a regular crystalline pattern. PVC is usually formed as a sort of *Glass*. Molecules nearby one another tend to be approximately aligned, but the alignments tend to alter slowly and randomly from one place to another in the solid. The material is a bit like a frozen liquid, or a liquid with a very high viscosity. The result is as if we had started to built a crystal, but kept changing our mind about where to put the layers of molecules. In any small region the groove wall may be quantised, but the details of the quantisation vary from place to place along the groove. For a recorded signal this produces an effect similar to *dithering* a signal before digital sampling. The randomised quantisation becomes indistinguishable from random noise. This dithering effect is enhanced by random thermal movements of the molecules. When playing an LP the effects of this molecular quantisation therefore appear as <u>noise</u>, not obvious quantisation distortions.

Another factor working in the LP's favour is that the stylus does not just touch the groove wall at a single point. Instead it presses against a finite *Contact Area*. This means that the force which positions the stylus is produced by a number of atoms in the groove surface. The contact area of a good stylus is typically the order of 10 μm square. Hence the stylus rests upon hundreds or thousands of PVC molecules at any time. The pressure of the stylus will tend to squeeze the groove surface. This makes it deform elastically until the total force exerted by all the displaced molecules is enough to support the stylus. Adding or removing a few PVC molecules in the contact area would shift the stylus by an amount which is much less than the size of a single molecule. The finite contact area of the stylus means that it essentially making a measurement which is averaged over many molecules. A larger contact area would permit the stylus to resolve smaller changes in the groove wall by averaging over more atoms. This averaging process, along with the physical dithering mentioned earlier, can let the stylus recover signal levels equivalent to changes in the groove

106

wall which are smaller than an individual molecule.

A time-varying output signal is obtained by drawing the stylus along the groove. Hence the frequency of a recorded signal variation is inversely proportional to its length along the groove. Since the stylus cannot be expected to respond to surface details which are much smaller than the width of its contact area, it follows that any improvement in resolution obtained by increasing the contact area may be purchased at the cost of a reduction in the available signal bandwidth. Alternately, we could choose a smaller stylus and sacrifice resolution for a wider bandwidth. The recorded signal is essentially both quantised and sampled by the atomic structure of the LP material, although in a way which varies from place to place on the disc.

High performance LP systems usually employ an *Elliptical* stylus (or some other near-equivalent). These styli are manufactured to have a specially shaped contact area which is shortened along the direction of travel and elongated perpendicular to it. The modified shape helps the stylus trace out higher frequencies (shorter groove wavelengths) without reducing the contact area. This improves the noise/bandwidth/distortion performance, but it can't entirely overcome the problems mentioned above. The stylus must have a non-zero contact area, hence the physical problems we've considered always apply.

It would be possible to go on considering various other factors which alter the detailed performance of Long Playing records. For example, any serious comparison of 'LP versus CD' would have to take into account the relatively high levels of signal distortion which commercial cartridges produce when recovering signals louder than the 0 dB level. Typically, signals of +10 dB or above are accompanied by harmonic distortion levels of 10% or more — not a very high fidelity performance! Even at the 0 dB level, most cartridges produce 1% or more harmonic distortion. The frequency response of signals recorded on LP are also modified — the high frequency level boosted and the low frequency level reduced — to obtain better S/N and distortion performance. This means that an LP replay system must include a *De-Emphasis* network to *Correct* the recovered signal's frequency response. Here, however, we are only interested in considering those physical factors which make the LP less than an ideally 'analog' way to communicate information. These extra factors affect the performance of an LP but they don't change the basic nature of the system.

107

The above analysis is a simplified one. It leaves out many features of a practical LP system. Despite that, it does serve to show that even a system which appears essentially 'analog' will still have underlying properties similar to a digital information processing system. In fact a similar situation arises with <u>all</u> analog signals in the real world since every physical process will be found to behave in a quantised manner when examined in sufficient detail. Despite this we do not usually observe any structured quantisation or sampling effects because they tend to be masked by a relatively high level of thermal noise and the averaging or smoothing effects of processes like the stylus's finite contact area. In effect, the real world beat us to the idea of using noise dithering to make quantisation effects invisible.

An argument similar to the one used to analyse the LP can be applied to sound waves themselves. The air consists of an enormous number of molecules whose sizes/shapes/energies/etc are quantised. The physical interactions between these molecules — i.e. they way in which they exchange energy and momentum with one another — follow the rules of quantum mechanics. Hence if we analyse sound waves in enough detail we should discover quantised behaviour once again. Just as with the LP groove, however, these effects are on such a small scale that we don't normally notice them. Usually we can describe sound in terms of the averaged statistical properties (pressures, mean velocities and displacements) of relatively large numbers of molecules without noticing this fact. This allows us to use the classical physics which describes sound in terms of continuous algebraic functions which satisfy a set of wave equations. Despite this, the individual molecules know nothing about our equations. The overall 'analog-like' properties of soundwaves arise because of the dithering/averaging effects of the countless individual quantised molecule–molecule interactions.

Summary

You should now understand that the terms 'analog' and 'digital' are based on idealisations. Real systems and signals will show a mixture of analog (smooth continuous) and digital (quantised) properties. Although it's often convenient to assume a signal/system is one thing or the other, this mixed behaviour is an unavoidable consequence of the way the world works.

Questions

1) A monophonic long-playing (LP) test record is being replayed using a cartridge (i.e. a transducer) whose *Sensitivity* $k = 0.2$ mV/cm/s. The recording is of a continuous 1 kHz sinewave tone whose level is +26 dB (referenced to a peak velocity of 5 cm/s). What is the rms value of the output signal voltage generated by the cartridge?

[14·1 mV rms.]

2) The test LP mentioned above is made of a material whose molecules average 10 nm in diameter. The +26 dB tone represents the highest signal level the transducer can produce without 'mistracking'. Assume that the LP material is crystalline and work out the system's *Dynamic Range* in dBs. How many bits-per-sample would be required for a digital system of the same bandwidth to provide the same dynamic range? Explain briefly why a non-crystalline material is a better choice for making LPs. **[Dynamic range = 90 dB. 15 bits per sample.]**

Chapter 13

Sensors and amplifiers

13.1 Basic properties of sensors

Sensors take a variety of forms, and perform a vast range of functions. When a scientist or engineer thinks of a sensor they usually imagine some device like a microphone, designed to respond to variations in air pressure and produce a corresponding electrical signal. In fact, many other types of sensor exist. For example, I am typing this text into a computer using an array of 'keys'. These are a set of pressure or movement sensors which respond to my touch with signals which trigger a computer into action. The keys respond to the pattern of my typing by producing a sequence of electronic signals which the computer can recognise. The information is converted from one form — finger movements — into another — electronic pulses.

Every sensor is a type of transducer, turning energy from one form into another. The microphone is a good example; it converts some of the input acoustical power falling upon it into electrical power. In principle, we can measure anything for which we can devise a suitable sensor. In this chapter we will concentrate on sensors whose output is in the form of an electrical signal which can be detected and boosted using an amplifier. However, similar results would be discovered if we examined sensors whose output took some other form such as water pressure variations in a pipe or changes in the light level passing along an optical fibre.

The basic properties of a sensor and amplifier are illustrated in figure 13.1. This shows an electronic sensor coupled to the input of an amplifier. Note that, so far as the amplifier is concerned, the sensor is a signal 'source' irrespective of where the signal may initially come from. The amplifier doesn't know anything about people singing into microphones or fingers bashing keyboards. It simply responds to a voltage/current presented to its input terminals.

The input to the sensor stimulates it into presenting a varying signal voltage, V_s, to the amplifier. The amplifier has an input resistance, R_{in}. (Both the source/sensor and the amplifier also have some capacitance, but for now we'll ignore that.) The signal power level entering the

110

amplifier's input will therefore be

$$P_{in} = \frac{V_s^2}{R_{in}} \qquad \text{...(13.1)}$$

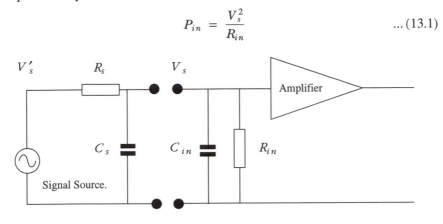

Figure 13.1   Source – amplifier combination.

Now P_{in} <u>must</u> be finite and limited by whatever physical process is driving the sensor. Yet equation 13.1 seems to imply that we could always get a higher power level from the source by changing to an amplifier with a lower *Input Resistance, R_{in}*. This apparent contradiction can be resolved by accepting that the voltage, V_s, seen coming from the source must, itself, depend upon the choice of R_{in}. The way in which this occurs should be clear from figure 13.1. The sensor itself must have a non-zero *Source Resistance, R_s*, which its output passes through. As a result the signal voltage at the amplifier's input will be

$$V_s = \frac{V_s' R_{in}}{(R_s + R_{in})} \qquad \text{...(13.2)}$$

where V_s' is the 'internal' voltage or *Electromotive Force (emf)* the sensor creates from the input which is driving it. The value of V_s' only depends on the input the sensor/transducer is responding to. It is unchanged by the choice of the amplifier, but the voltage seen by the amplifier depends upon the source and amplifier resistances so the power entering the amplifier will be

$$P_{in} = \frac{V_s'^2 R_{in}}{(R_s + R_{in})^2} \qquad \text{...(13.3)}$$

In order to maximise the signal power entering the amplifier we should arrange that $R_{in} = R_s$. A lower input resistance would *load* the source too much, causing V_s to fall. A higher input resistance would reduce the current set up by the signal voltage. In effect, making the source and amplifier resistance values the same means we can get the biggest possible

111

voltage–current product at the amplifier's input. Since power = voltage × current this ensures the highest possible input power for a given signal emf, V'_s. This result is a general one which arises because the amount of power generated by a source can never be infinite. All signal sources will have a non-zero source resistance (or *Output Resistance*). In a similar way we can expect all real amplifiers and signal sources to exhibit a non-zero capacitance. This is called the *Source Capacitance* for a source/sensor and the *Input Capacitance* for an amplifier.

From figure 13.1 we can see that these two capacitances, C_s and C_{in}, are in parallel. For the voltage seen at the amplifier's input to be able to change we have to alter the amounts of charge stored in these capacitances. The current required to do this must come through R_s and R_{in}. From the point of view of the capacitors these offer two parallel routes for charge to move from one end of the capacitors to the other — i.e. they appear in parallel. This combination of capacitance and resistance means that the voltage V_s, seen by the amplifier cannot respond instantly to a swift change in the source voltage, V_s'. Changes in V_s are 'smoothed out' with a time constant, $\tau = RC$, where R and C are the parallel combinations of the input and amplifier values.

In some cases these resistances and capacitances are actual components put in the system. In other cases they are a result of some other physical mechanisms. In each case their effects can be modelled using the kind of circuit shown in figure 13.1. Irrespective of whether they're deliberate additions or 'stray' effects, these capacitances and resistances are always non-zero. Hence it is impossible to change a measured signal level infinitely quickly. This is another way of stating the basic principle of information processing that no signal can have an infinite bandwidth (i.e. reach infinite frequencies). If it did, it would be able to convey an infinite amount of information in a limited time. Alas, in the real world this is impossible.

13.2 Amplifier noise

When designing or choosing a measurement system we need to be able to compare the performances of various amplifiers to select the ones most appropriate for the job in hand. Various criteria affect the choice, ranging from price to gain. When making accurate measurements it is usually preferable to choose amplifiers which generate the lowest noise level.

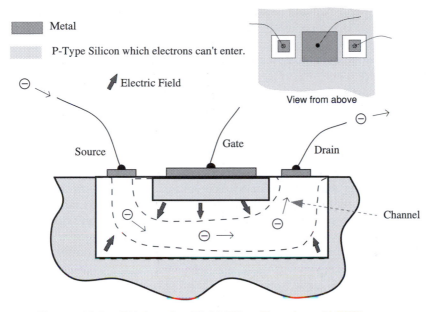

Figure 13.2 PN Junction Field Effect Transistor (J-FET).

A wide range of devices have been used to amplify signals. Although their details differ we can expect that they will operate at a temperature above absolute zero and, as a result, must produce some thermal noise. Similarly, for their input and output signals to have non-zero powers, they must pass some current, hence producing some shot noise. It seems to be one of the basic laws of Nature (Murphy's Law?) that all gain devices, from MOSFETs to valves, generate *Excess* noise — i.e. they all produce more noise than we would predict from adding together the thermal noise and shot noise. For the sake of example we can consider the behaviour of a *Field Effect Transistor* (FET) amplifier of the sort illustrated in Figure 13.2. The device shown is a simple *N-channel junction FET*. This is made by forming a channel of N-type semiconductor in a substrate of P-type semiconductor. The channel–substrate boundary forms a *PN junction* which behaves like a normal diode. As a result, provided we avoid forward biassing the gate–channel boundary:

• Almost no current flows between gate and channel
• The charge in the gate (and substrate) repels the free electrons in the channel and prevents them from coming too close to the walls of the channel. This produces *Depletion Zones* near the walls whose size depends upon the applied gate potential.

When we apply a voltage between the *Source* and *Drain* contacts, electrons flow through that part of the channel which has not been depleted. We can think of the channel as a slab of resistive material of length, L, and cross sectional area, A. For a material of resistivity, ρ, such a slab would have an end-to-end resistance, $R = \rho L / A$. Varying the gate voltage alters the depletion zones and hence changes the effective cross sectional area, A, of the channel. As a consequence, when we vary the gate potential the effective resistance between source and drain changes. The FET therefore acts as a source–drain resistor whose value depends upon the gate potential. This description of the operation of an FET is too simple to explain all the detailed behaviour of a real device but it's OK for many purposes. In practice the drain-source voltage is usually sufficiently large that the potential difference between the drain and gate is much greater than that between source and gate. As a result the depletion region inside the channel is much smaller at the source end than at the drain — i.e. the cross-sectional area of the effective channel is quite thin at one end.

From the simple description given above we would expect the channel current to increase in proportion with the applied drain–source voltage. However there is a tendency for any increase in drain voltage to enlarge the depleted region near the drain. This reduces the channel area, limiting any current increase. As a result we find that, for reasonably large drain–source voltages, the FET behaves more like a device which passes a drain–source current controlled by the gate potential. Because of this the gain of an FET is usually given in terms of a *Transconductance*. This can be defined as the change in drain–source current divided by the change in gate potential which causes it.

The gate-channel is normally reverse biassed, so almost no gate current is required to maintain a given gate potential. As a consequence the input resistance of an FET is very high, typically 10 MΩ or more. However, to alter the gate potential we must vary the charge density within the gate. This means that we have to move some charge into or out of the gate. As a consequence the gate–channel junction has a small capacitance. For a typical FET the gate–channel capacitance is a few tens of pF or more.

Noise is generated within the FET by various physical processes. For example:

    i) Shot noise fluctuations in the current flowing through the channel

    ii) Thermal noise in the channel resistance

    iii) Thermal motions of the gate charge carriers, producing random

114

fluctuations in the size and shape of the depletion region — and hence in the channel resistance.

All of these effects (and others which have been ignored) will vary according to the bias voltages and currents, details of the semiconductor doping, device geometry, and temperature. Instead of risking becoming bogged down in a detailed analysis of these effects (which may be futile as some of the underlying processes are poorly understood!) we can model the behaviour of the FET (or any other gain device) in terms of a fictitious pair of *Noise Generators*. This approach is very useful when we are mainly concerned with comparing one amplifier with another and don't want to bother with the details of where the noise is actually coming from.

Figure 13.3a represents a simple amplifier using an FET. The noise produced by the real FET and the other components which make up the amplifier are assumed to come from a mythical *Noise Voltage Generator*, e_n, and *Noise Current Generator*, i_n, connected to the amplifier's signal input. Figure 13.3b represents the way in which this idea can be generalised to apply to any amplifier, irrespective of its design. The noise performance of any amplifier can now be described by the appropriate values of e_n and i_n. These are normally specified as an rms voltage and current *spectral density* — the units of e_n usually being $nV/\sqrt{Hz}$, and i_n $pA/\sqrt{Hz}$. Figure 13.4 illustrates the typical manner in which they vary with fluctuation frequency.

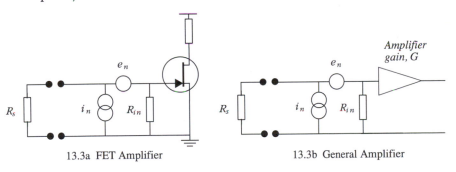

13.3a FET Amplifier   13.3b General Amplifier

Figure 13.3   Noise models of amplifers.

A noise producing process which has not been mentioned in previous chapters is *Generation-Recombination Noise* (GR-noise). A large number of electrons do not normally take part in conduction as they do not have enough energy to escape their orbit around a particular atom. Every now and then, however, one of these *Bound* electrons may interact with a passing electron or a lattice vibration (i.e. a phonon) and gain enough

115

energy to escape. This process can be regarded as 'lifting' an electron up into the conduction band and leaving behind it a 'hole' in a lower band.

Sometimes the newly freed electron does not move away swiftly enough to avoid dropping back into the hole. But if it manages to get away we find that a pair of extra charge carriers have joined those able to provide current flow through the material. Eventually, an electron will pass close enough to the hole to fall into it and the total number of available charge carriers will return to its original value. This process means that the current flowing through the channel as a result of an applied voltage will tend to fluctuate. (Note that this process is different from shot noise.)

There is a difference in potential between the channel and the gate/substrate. Any new electron–hole pairs generated near the channel walls will tend to be pulled apart. For an N-channel FET the field will sweep the 'new' electron into the channel and pull the hole back into the substrate. As a consequence, the random creation of carrier pairs in the region near the gate–channel junction produces a small, randomly varying, current flowing across the boundary. This in turn causes random variations in the size and shape of the depletion region which produces an extra noise current in the channel.

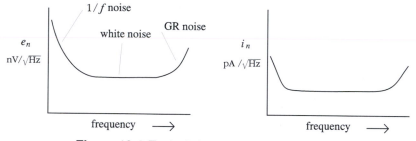

Figure 13.4 Typical shapes of noise power density spectra of noise generators.

From figure 13.4 it can be seen that, at high frequencies, the noise power spectral density tends to increase with frequency. This is due to GR-noise produced by quantum mechanical effects. Although energy must be conserved overall, quantum mechanics permits the energy of a system to fluctuate by an amount ΔE provided the fluctuation only lasts a time $\Delta t \approx h / \Delta E$ (h being Planck's constant). In a semiconductor whose energy gap is ΔE this means that electron–hole pairs may be created without the required specific energy input, ΔE, provided they vanish again in a time $\Delta t \approx h / \Delta E$. As a result, when we consider periods of time

116

which are less than this time the density of carriers in the material appears to fluctuate randomly.

These short-lived random variations in the number of free charges mean that the current which flows in response to an electric field also varies. If we consider shorter periods we are allowed to consider larger energy fluctuations and an increasing number of electrons, tied more strongly to their atoms, can briefly join in this process. Hence this effect produces a noise level which increases with frequency (i.e. with decreasing fluctuation period). This effect does not create noise power out of nothing. The initial ΔE is a sort of 'loan' which must be repaid since, if we want to observe a change in the current, we must apply an electric field to drag the electron–hole pair apart. This field hence does some work in producing the extra current.

All gain devices exhibit some amounts of voltage noise, e_n, and current noise, i_n. The precise levels they produce — and their frequency spectrum — depends upon the type of device, how it is made and operated. When comparing bipolar transistors with FETs we generally find that bipolar devices have higher current noise levels and FETs have higher voltage noise.

13.3 Specifying amplifier noise

In practice we are often not told how e_n and i_n vary with frequency for a particular amplifier. Instead we are presented with a single value which indicates the overall amount of noise the amplifier produces. This value may be specified in various ways. The most common measures are the *Noise Resistance*, R_n, the *Noise Temperature*, T_n, the *Noise Factor*, F, and the *Noise Figure*, M. Whilst any one of these values can be useful for encapsulating the behaviour of an amplifier it should be clear that a single number cannot contain all the information offered by a detailed knowledge of the e_n and i_n spectra. They should therefore be used with care.

Figure 13.5 illustrates a system which amplifies the signal voltage, v_s, generated by a source whose output resistance is R_S. The amplifier is assumed to have a voltage gain, A_V, input impedance, R_{in}, and produces a noise level equivalent to a combination of a noise voltage generator, e_n, and noise current generator, i_n, located as shown at the amplifier's input. A signal source at a temperature, T, will itself produce thermal noise

117

equivalent to a voltage generator whose rms magnitude is

$$e_s = \sqrt{4kTBR_s}$$ ... (13.4)

placed in series with the source.

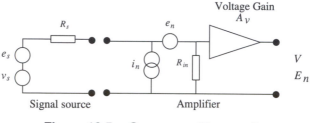

Figure 13.5   Source-amplifier coupling.

For the sake of simplicity we can assume a unit bandwidth ($B = 1$ Hz) and that the source does not produce any other form of noise. This means that the source is as 'noise-free' as we can expect in practice. Taking into account all of the noise generators shown in figure 13.5, the total rms noise voltage, E_0, which is <u>output</u> by the amplifier will be such that

$$E_0^2 = |A_V|^2 \cdot \left[\left\{ \frac{e_n R_{in}}{R_s + R_{in}} \right\}^2 + \left\{ \frac{e_s R_{in}}{R_s + R_{in}} \right\}^2 + \left\{ \frac{i_n R_s R_{in}}{R_s + R_{in}} \right\}^2 \right]$$ ... (13.5)

and the source signal, v_s, will produce a voltage

$$V_0 = \frac{A_V R_{in} v_s}{R_s + R_{in}}$$ ... (13.6)

at the amplifier's output. We can now define the <u>system</u> gain, H, (as distinct from the amplifier gain, A_V) as

$$H \equiv \frac{V_0}{v_s} = \frac{A_V R_{in}}{R_s + R_{in}}$$ ... (13.7)

Note that this value takes into account both the amplifier's voltage gain and the voltage attenuation produced by R_s and R_{in} acting as a potential divider (attenuator) arrangement. Hence this gain will always be smaller than A_v.

We can now regard the <u>total</u> noise at the output of the system as being due to a single voltage generator, e_t, which replaces e_s. From the above definition of the system gain we can expect that

$$e_t = \frac{E_0}{H}$$ ... (13.8)

which, combining the above expressions, leads to the result

Sensors and amplifiers

$$e_t^2 = e_s^2 + e_n^2 + i_n^2 R_s^2 \qquad \ldots (13.9)$$

The noise in the system has now been gathered into a single number, e_t, whose value indicates the total noise present in the system. From this we can define each of the noise measures mentioned earlier.

The *Noise Factor, F,* is defined as

$$F \equiv \text{(total noise power)} / \text{(source resistance noise power)}$$

i.e.

$$F = \frac{e_t^2}{e_s^2} = \frac{e_s^2 + e_n^2 + i_n^2 R_s^2}{e_s^2} \qquad \ldots (13.10)$$

The *Noise Figure, M* is defined to be the noise figure quoted in decibels

$$M \equiv 10. \, \text{Log}\{F\} \qquad \ldots (13.11)$$

For a perfectly noise-free amplifier e_n and i_n would both be zero. Such an amplifier would have a noise factor of unity and a noise figure of 0 dB.

The *Noise Resistance, R_n,* can be defined by equating the amplifier's contribution to the total noise to a thermal noise level

$$4kT R_n \equiv e_n^2 + i_n^2 R_s^2 \qquad \ldots (13.12)$$

where T is taken as the physical temperature of the amplifier (normally assumed to be around 300 K).

Because of the possibility of confusing the amplifier's noise resistance with its input resistance it is prudent to avoid the use of noise resistance values.

The *Noise Temperature, T_n,* defined by

$$4kT_n R_s \equiv e_n^2 + i_n^2 R_s^2 \qquad \ldots (13.13)$$

is a more acceptable alternative since it avoids this confusion. Note, however, that this temperature value is <u>not</u> the physical temperature of the amplifier!

When comparing amplifiers and gain devices listed in manufacturer's catalogues we're frequently only given one of the above measures as an indication of the noise level. When examining these figures it is important to compare like with like. All of the above measures explicitly depend upon the chosen source resistance, R_s. Furthermore, the frequency dependence of e_n and i_n will vary from one gain device to another. As a result two values of a noise measure are not directly comparable if they are given for different frequencies.

119

To measure the voltage and current noise levels of a particular amplifier we can observe the effects of short-circuiting and open-circuiting the amplifier input terminals (i.e. setting R_s to zero and to infinity). When R_s = 0 the current noise present cannot produce any observable voltage. The output noise from an amplifier whose input is shorted is therefore due only to its input voltage noise generator, e_n.

When we open-circuit the amplifier input we produce an effective source resistance of $R_s = \infty$. The noise current generator now produces an rms voltage $i_n R_{in}$ across the amplifier's input resistance. The noise fluctuations this produces are uncorrelated with those produced by the noise voltage generator. Hence they combine to produce a total rms noise voltage at the amplifiers input of $\sqrt{e_n^2 + i_n^2 R_{in}^2}$ when the amplifier input is open-circuit. By measuring the amplifier's output noise level in both situations we can therefore determine values for both e_n and i_n.

Summary

You should now know that all signal sources must have a non-zero *Source* (or *Output*) *Resistance* and a non-zero *Source Capacitance*. That all the noise mechanisms in a system can be simplified into a an equivalent pair of mythical *Noise Generators* at the input to the system. A 'new' noise mechanism, *Generation-Recombination* has been introduced and it's power spectral density has been seen to increase with fluctuation frequency.

You should also now know that the total system noise can be simplified into a single generator value and the result may be specified in terms of various figures — *Noise Temperature, Resistance, Figure,* or *Factor.* It should also be clear that a single figure of this kind can only be used to compare one amplifier to another when the source resistances are the same. You should also now know that the current and voltage noise levels of an amplifier can be measured by recording the output noise level when the amplifier's input is open- and short-circuited.

Questions

1) Explain why we can transfer the maximum possible signal power from source to load when the source and load resistances have the same value.

2) An amplifier has an input resistance of R_{in} = 50 kΩ, and its noise behaviour can be defined in terms of voltage generator and current generator *Noise Spectral Densities* of e_n = 5×10^{-9} V/$\sqrt{\text{Hz}}$ and i_n = 10^{-12} A/$\sqrt{\text{Hz}}$ respectively. A sensor whose source resistance is 22 kΩ is connected to the amplifier's input. The sensor is at 300 K and only generates thermal noise. What is the value of the system's *Noise Factor*. What is the value of the system's *Noise Temperature*? [**F = 2·39. T_n = 419 °K.**]

3) Explain how you can measure the values of an amplifier's effective noise voltage and current generators.

Chapter 14

Power coupling and optimum S/N

14.1 Optimising signal to noise ratio

Sometimes we can alter the physical details of a signal source or use a transformer to change the source's apparent resistance whilst maintaining the available signal power. In earlier chapters we found that the amount of noise and signal power we see coming from a system depends upon the source resistance. This raises the question — is there a value for the source resistance which produces an optimum (i.e. maximum) signal to noise ratio? If so, what is this value?

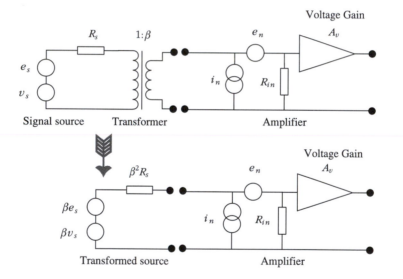

Figure 14.1   Source–Amplifier coupling and power transformation.

Figure 14.1 illustrates the use of an idealised transformer which has a turns ratio of $1:\beta$. This steps up/down the output signal and noise voltages produced by the source to βv_s and βe_s, respectively. The transformer cannot output any more power than it receives. For an ideal (loss–free) transformer the input and output powers will be the same. As the output voltage is a factor β times that generated by the source it follows that the output current must be $1/\beta$ times that flowing through the source.

Consequently, the combination of the source and transformer appears to any following circuit to have an effective source resistance of $R_s' = \beta^2 R_s$.

Using the same argument as in the previous chapter we can say that the total noise level for the system is equivalent to an rms voltage, e_t, such that

$$e_t^2 = (\beta e_s)^2 + e_n^2 + (i_n \beta^2 R_s)^2 \qquad \dots (14.1)$$

A signal voltage, v_s, generated by the source will produce a signal/noise ratio

$$S/N = \frac{(\beta v_s)^2}{e_t^2} \qquad \dots (14.2)$$

Clearly, this depends upon the choice of β. Whenever possible it would be preferable to select the value of β which maximises S/N. This is equivalent to the value which <u>minimises</u> $e_t^2 / (\beta v_s)^2$. The optimum choice of β can therefore be found from

$$\frac{d}{d\beta}\left[\frac{e_t^2}{(\beta v_s)^2}\right] = 0 \qquad \dots (14.3)$$

i.e.

$$\frac{2\beta i_n^2 R_s^2}{v_s^2} - \frac{2e_n^2}{\beta^3 v_s^2} = 0 \qquad \dots (14.4)$$

which is satisfied when

$$\beta^2 = \frac{e_n}{i_n R_s} \qquad \dots (14.5)$$

Since the transformed source resistance, R_s', presented to the amplifier is $\beta^2 R_s$ it follows that the optimum value for this resistance will be

$$R_s' = \frac{e_n}{i_n} \qquad \dots (14.6)$$

For the above argument it was assumed that the source resistance presented to the amplifier could be altered using a transformer. In some other situations we can modify the signal source or replace it with another and alter the source resistance without altering the available signal power. Irrespective of how this is done the above result tells us that — for an amplifier whose noise is represented by a voltage generator, e_n, and current generator, i_n — the maximum possible signal/noise ratio will be obtained when the source resistance equals e_n / i_n.

In the last chapter we saw that the optimum signal power transfer will occur when we choose a source resistance which equals the amplifier's input resistance. In general, e_n / i_n does <u>not</u> equal the input resistance of the amplifier. As a result, the source resistance which provides the best signal power transfer usually <u>isn't</u> usually the value which gives the best possible S/N ratio.

Books on electronics tend to recommend that, whenever possible, we arrange that the source's output resistance and the amplifier's input resistance should be *matched* — i.e. have the same value. (The same approach is recommended when the signal is carried using a *transmission line.*) This gives the most efficient transfer of signal power, but may result in a S/N ratio below the highest possible value. As a result there is often a conflict and we have to choose <u>either</u> a source resistance which provides the highest possible signal/noise ratio <u>or</u> a source resistance which maximises the signal power transferred.

In practice we are often presented with a source whose properties are fixed, but we can select which amplifier to use from an available range. Each amplifier has a particular input resistance, R_{in}, and a noise level equivalent to specific e_n and i_n values. Because of the conflict between optimum signal transfer and signal/noise ratio there won't usually be a 'perfect' choice of amplifier — unless we're lucky enough to find one where $R_s = R_{in} = e_n / i_n$. Instead, we must usually make a choice based upon an assessment of the relative importance of these factors for the job in hand.

14.2 Behaviour of cascaded amplifiers and transmission lines

Figure 14.2 illustrates the use of a pair of amplifiers to increase the signal from a source. Each amplifier has an input resistance, Z_I, and an output resistance, Z_O.

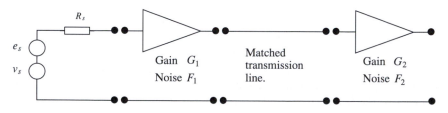

Figure 14.2   Cascaded amplifiers and connection.

The signal power input into the first amplifier will be

$$P_I = \frac{V_S^2 Z_I^2}{(R_S + Z_I)^2} \qquad \ldots(14.7)$$

For a given signal voltage this power will be maximised if $R_S = Z_I$. As a result we should usually try to equate (match) these impedances whenever it is convenient to do so. A similar result arises when we connect amplifiers together. Each amplifier views the preceding one as a source having a particular resistance. Whenever possible we should arrange that the input impedance of an amplifier should equal the output impedance of the preceding one. This ensures that signal power is not wasted.

In the arrangement shown in figure 14.2 the amplifiers are connected by a length of transmission line of *Characteristic Impedance*, Z_C. Co-axial cables, pairs of wires, microwave waveguides, light fibres, etc, are all examples of transmission lines. Each can be used to carry signals over long distances. To understand the concept of characteristic impedance, imagine a signal source transmitting a signal into an infinitely long transmission line. To transmit power along a line it has to send both a non-zero voltage (or electric field) and a non-zero current (or magnetic field) out along the line. This power then moves away from the source, along the infinitely long cable, never to return.

The amount of current the source has to put into the cable to 'drive' a given voltage will depend upon the type of transmission line. However, so far as the source is concerned, the power transmitted into the cable is 'lost' just as if the cable were a resistor. The value of the resistor which would require the same voltage/current ratio is said to be the characteristic impedance of the line. If we end a <u>finite</u> length of line with a load whose resistance equals the line's characteristic impedance the current/voltage ratio of the signal perfectly matches that required by the load. Hence all the signal power flows into the load.

In figure 14.2 the load at the output end of the line is the input impedance, Z_{IN}, of the second amplifier. By arranging that $Z_{IN} = Z_C$ we ensure that all the signal power passing along the line is coupled into the second amplifier (this assumes, of course, that the transmission line doesn't lose any of the power on the way!). So far as the first amplifier is concerned, it then sees an output load resistance, Z_C, since none of the power it transmits comes back to it. As a result, to efficiently transmit signal power along the transmission line we should try to arrange that $Z_{IN} = Z_C = Z_{OUT}$.

125

We'll assume the impedances throughout the system <u>have</u> been matched — although from the previous section it should be noted that this may not give the highest possible signal/noise ratio. The first amplifier has, when matched, a noise factor, F_1, and a power gain, G_1. The second has, when matched, a noise factor, F_2, and power gain, G_2.

Thermal noise in the source will produce a noise power spectral density at the input of the first amplifier of

$$N_1 = \frac{e_s^2}{4R_S} = kT \qquad \qquad \text{... (14.8)}$$

From the definition of noise factor, it follows that this amplifier supplies the following one with a noise power spectral density of

$$N_{01} = F_1 G_1 kT \qquad \qquad \text{... (14.9)}$$

If we were to connect the second amplifier's input directly to a matched source resistance instead of linking it to the output from the first amplifier it would supply an output noise power per Hertz bandwidth

$$N_{02} = F_2 G_2 kT \qquad \qquad \text{... (14.10)}$$

Since the source resistance would itself be generating a noise power spectral density, kT, an amount $G_2 kT$ of what we see coming out of the amplifier would originate in the <u>source</u>, not the amplifier. The noise power per hertz bandwidth which is generated <u>inside the second amplifier</u> is therefore $F_2 G_2 kT - G_2 kT$.

The total output noise power spectral density for the arrangement in figure 14.2 will therefore be

$$N_T = G_2 (F_1 G_1 kT) + G_2 (F_2 - 1) kT \qquad \qquad \text{... (14.11)}$$

We can consider the combination of amplifiers as a single 'multi-stage' amplifier whose power gain is $G_1 G_2$. This combination can then be defined to have an overall noise factor, F_T, such that

$$N_T = F_T G_1 G_2 kT \qquad \qquad \text{... (14.12)}$$

Amplifiers connected in this way are said to be *Cascaded* or chained together. Combining the above expressions we find that the noise factor of the two cascaded amplifiers will be

$$F_T = F_1 + \frac{F_2 - 1}{G_1} \qquad \qquad \text{... (14.13)}$$

When using an initial amplifier whose gain, G_1, is moderately high this result implies that — unless F_2 is very large compared with F_1 — the cascaded pair has a noise factor, $F_T \approx F_1$. Consider, as an example, a case where $F_1 = 1 \cdot 5$, $F_2 = 2$, and $G_1 = 100$. Using expression 14.13 we can calculate that the cascaded amplifiers will have an overall noise factor of $F_T = 1.501$, i.e. even though the second amplifier is relatively noisy the overall system's noise factor is almost entirely due to the first amplifier.

This result arises because the signal level presented to the second amplifier is much larger than that presented to the first. Hence the second amplifier would have to generate a considerable amount of noise to significantly degrade the overall signal/noise ratio. For this reason we usually only need to ensure that the first amplifier in a chain has a low noise factor. However, it should be noted that this may not be true if the transmission line which connects the two amplifiers is imperfect.

Any real transmission line will lose some of the signal power it is given to convey. For example, a co-axial cable will dissipate some power due to the resistance of its metal conductors. The transmission line will change (attenuate) the signal power by a factor, a, i.e. an output power, P, supplied by the first amplifier will provide a power, aP (where $a \leqslant 1$), to the second.

In many cases a will be close to unity. Under these circumstances the combination of the first amplifier and transmission line have an overall power gain, aG_1, and we need only worry about the first noise factor, F_1. However, if the transmission line is long enough and a is low enough for aG_1 to become comparable with (or less than!) unity, we find that the signal power reaching the second amplifier is <u>not</u> significantly larger than that reaching the first. Under these circumstances the noise factors of both amplifiers become important.

The above argument for two cascaded amplifiers can be extended to situations where three or more are chained together. For example, for three amplifiers in a chain the overall noise factor (neglecting transmission line losses) would be

$$F_T = F_1 + \frac{F_2 - 1}{G_1} + \frac{F_3 - 1}{G_1 G_2} \qquad \text{...(14.14)}$$

Summary

You should now know how the S/N ratio we can obtain from a signal source depends upon the choice of source resistance. It should also be clear that — in general — the best possible S/N ratio requires a <u>different</u> source/amplifier resistance than the *Matched* values which maximise the power transfer. You should now know how the noise performance of a *Cascade* of amplifiers and connections depends upon their gain and noise performance. In most practical cases it is sensible to use a 'low noise preamp' to boost a signal being fed to later amps whose noise performance is less important.

Questions

1) A source of resistance, R_s, is connected to an amplifier whose input resistance is R_{in} via a transformer which has a *Turns Ratio* of 1:β. The amplifier's noise is specified in terms of a pair of noise voltage and current generators, e_n and i_n. Derive an expression for the value of β which provides the highest possible signal-to-noise ratio.

2) An amplifier has an R_{in} = 100 kΩ, e_n = $4{\times}10^{-9}$ V/$\sqrt{\text{Hz}}$, i_n = 10^{-13} A/$\sqrt{\text{Hz}}$. It is connected to a 10 kΩ source via a transformer. What transformer's turns ratio value would provide the highest signal to noise ratio? What would ratio would provide the greatest signal power transfer? [**Best S/N from 1:2. Best signal power 1:3·16.**]

3) The amplifier described in question 2) has a voltage gain, A_v = 1000. It is connected to the source via a transformer which provides the optimum signal to noise ratio. Assuming that the source noise is purely thermal and its temperature is 300 K, what is the value of the noise power spectral density (in microvolts per root Hertz) at the amplifier's output? (Hint, look at section 13.3 again.) [**18 μV/$\sqrt{\text{Hz}}$**]

4) A signal is amplified by a cascade of two amplifiers. The impedances throughout the system are *Matched*. The first amplifier has a power gain of G_1 = 10 and a noise factor, F_1 = 1·1. The second has a power gain G_2 = 1000 and a noise factor F_2 = 2·5. What is the value of the cascade's total noise factor? [**1·25**]

Chapter 15

Signal averaging

15.1 Measuring signals in the presence of noise

When measuring a small but steady signal in the presence of random noise we can often improve the accuracy of the result by making a number of measurements and taking their average. This approach has the great advantage that it is easy to do — given enough time — but it cannot overcome all of the practical problems which arise when making real measurements. In particular, there are two sorts of problem which simple averaging copes with rather poorly: *1/f-noise*, and the presence of *Background* effects.

When considering the merits of various signal processing systems we're primarily interested in comparing the signal/noise ratios they can offer. It's this ratio which largely determines how precise a measurement can be. A low signal level can always be enlarged if we can afford a suitable amplifier. However, this won't lead to a more accurate result if the measurement was already noise limited because we'll boost the noise level along with the signal.

Note that the following arguments assume the power gain, G, of an amplifier (or filter) is simply equal to $|A|^2$ where A is the voltage gain. This is only really true when the amplifier's input resistance is equal to the output load resistance it drives. Similarly, it is assumed that the power, P, at any point is simply equal to $|V|^2$, where V is the rms signal voltage. This is only correct for a load resistance of unity (one Ohm). These assumptions make some of the mathematical expressions a bit simpler and don't change any of the conclusions. In practice, when working out the properties of a real system these factors have to be taken into account.

15.2 Problems of simple averaging

To illustrate these problems, consider the system shown in figure 15.1. A source, S, produces a response from a detector which is then amplified, and passed through an analog *Integrator* to a voltmeter. The integrator is made using an operational amplifier, resistor, and capacitor.

129

A normal operational amplifier has two signal input terminals, generally called the *Inverting* and *Non-Inverting* inputs (shown by the '–' and '+' signs on the diagram). The output voltage the op-amp produces is proportional to the <u>difference</u> between these two input levels. This arrangement allows the op-amp to be used as the heart of a *Feedback* arrangement. The voltage gain of a typical op-amp is very large (usually over 100,000) so a reasonable output voltage only arises when the voltages at the inverting (–) and non-inverting (+) inputs are almost identical. For example, if the output voltage is 1 V and the gain is 100,000 then the two inputs will only differ by 10 μV.

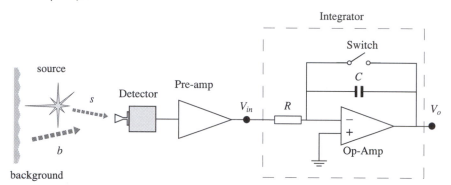

Figure 15.1   Analog integrator used to collect detected signal level.

In the circuit shown in figure 15.1 the non-inverting (+) input is connected directly to 0 Volts. The inverting input (–) is connected via a capacitor to the amplifier's output. The simplest possible state of this arrangement is when both input voltages, and the output voltage, are all at 0 V. We can therefore imagine the system starting off in this state.

When we apply an input voltage, V_{in}, to the resistor a current, $I = V_{in} / R$, will begin to flow through it as the other end of the resistor is initially at 0 V. This current starts flowing into the amplifier, stimulating a change in its output voltage. Because the signal is being presented to the inverting input the output voltage this produces will have the opposite sign to the input.

Any change in the output voltage will have to alter the amount of charge in the capacitor, C — i.e. a current will be drawn through the capacitor. As a result we find that most of the current flowing through the resistor passes on through the capacitor as the output voltage changes. Since the op-amp's gain is very large only a relatively tiny amount of the input current needs to actually enter the op-amp to generate the output voltage this process requires.

The small current, i, flowing into the op-amp's input will be the difference between the input and capacitor currents

$$i = \frac{V_{in}}{R} + C\frac{dV_O}{dt} \qquad \qquad ...(15.1)$$

As the amplifier gain is large we can expect that $i \ll \frac{V_{in}}{R}$ so we can reasonably assume that it is virtually zero and re-arrange 15.1 as

$$\frac{dV_O}{dt} = \frac{-V_{IN}}{RC} \qquad \qquad ...(15.2)$$

Having begin with an output voltage, $V_O = 0$, at a time, $t = 0$, we can therefore say that the output voltage at some later time, $t = T$, will be

$$V_O\{T\} = \int_0^T \frac{-V_{IN}}{\tau}\,dt \qquad \qquad ...(15.3)$$

where $\tau \equiv RC$ has the units of time and is called the *Time Constant* of the integrator. In effect, the system behaves as if all of the input current, I, is collected into the capacitor and the arrangement functions as an integrator, the output voltage being proportional to the time-integral of the input.

In practice the capacitor can be initially shorted by closing the switch connected across it. This sets the output voltage to zero. When a measurement commences the switch is opened and integration begins. For a steady input signal voltage, v, the output voltage after a time, T, will simply be proportional to vT. Hence the integrator performs the useful function of 'adding up' the signal voltage, v, over a period of time. As a result we need not actually take a series of voltage readings and calculate their average. Instead we can use an integrator, read V_O after a time, T, and define the average input signal voltage, $\langle v \rangle$, during this period to be

$$\langle v \rangle = \frac{-V_O\tau}{T} \qquad \qquad ...(15.4)$$

Any real integrator will be built using an op-amp powered from voltage rails which supply some specific fixed voltages. As a result, we cannot allow the circuit to go on integrating a signal voltage for an indefinite time as, eventually, V_O will reach the rail voltage and integration must then stop. To overcome this problem we may repeatedly read the output voltage, V_O, after a moderate time interval, t, and *reset* the integrator output to zero by briefly closing the shorting switch before allowing another integration

131

over another period, t. The resulting set of readings for V_O can then be added together to obtain the voltage which would have been reached if the circuit had been able to integrate successfully over the whole period. Many practical systems combine the use of an analog integrator with this method of repeated reading and resetting.

The effect of noise on an integrated result can be understood in terms of the integrator's effective *Power Gain* at any frequency, f. At any frequency the noise can be represented by a 'typical' input of the form

$$V_N = A_C \, \text{Cos}\{2\pi f t\} + A_S \, \text{Sin}\{2\pi f t\} \qquad \qquad \dots(15.5)$$

For real noise the values of A_C and A_S will vary randomly from moment to moment. This is because the phase of the signal is unpredictable. Their values at any instant are therefore independent, i.e. we can't predict one from knowing the other. However, on average, we can expect their magnitudes to be the same. We can therefore say that the time averaged power of this 'noise like' input will be

$$P_{in} = \frac{\langle A_C \rangle^2}{2} + \frac{\langle A_S \rangle^2}{2} = A^2 \qquad \qquad \dots(15.6)$$

where expression 15.6 essentially defines A to be the mean amplitude of each individual component. The factors of $1/2$ appear because we are averaging sin^2 quantities over a number of cycles.

Since the actual amplitudes of the sine and cosine components of the noise are statistically independent we can expect their contributions to the noise level at the integrator's output to also be independent. Their combined effect at the output will therefore equal the sum of the powers they individually produce. Integrating the effects of the two contributions over a period, T, we obtain two voltages. These must then be squared separately and then added to obtain the total output noise level

$$P_{out} = \left[\frac{1}{\tau} \int_0^T A \, \text{Cos}\{2\pi f t\} \, dt \right]^2 + \left[\frac{1}{\tau} \int_0^T A \, \text{Sin}\{2\pi f t\} \, dt \right]^2$$

$$= \frac{A^2 \, \text{Sin}^2\{\pi f T\}}{(\pi f \tau)^2} \qquad \qquad \dots(15.7)$$

We may define the integrator's power gain to be the ratio, $G \equiv P_{out}/P_{in}$. Comparing expressions 15.6 and 15.7 we can therefore say that

$$G\{f\} = \frac{\text{Sin}^2\{\pi f T\}}{(\pi f \tau)^2} \qquad \qquad \dots(15.8)$$

Having discovered the integrator's power gain we can now say that the total output power produced, after integration, by an input white noise power density, S, will be

$$N = \int_0^\infty S\,G\,\{f\}\,df = \int_0^\infty \frac{S\,\text{Sin}^2\{\pi f T\}}{(\pi f \tau)^2} = \frac{S\,T}{2\tau^2} \qquad \ldots(15.9)$$

The output signal power produced by integrating a steady input level, v, over a period, T, will be

$$P_s = V_O^2 = \frac{v^2 T^2}{\tau^2} \qquad \ldots(15.10)$$

Combining this with the result for noise we can therefore say that, when accompanied by an input 'white' noise power spectral density, S, we obtain a final signal to noise ratio of

$$\frac{P_s}{N} = \frac{2v^2 T}{S} \qquad \ldots(15.11)$$

This result is a very important one. It tells us that the signal to noise ratio of a measurement obtained using an integration method can increase linearly with the integration time, T. In practice this means we can often expect to improve the accuracy of a measurement by integrating for longer. The integration process is mathematically equivalent to making a series of measurements and adding them together. We can therefore generalise this result. If we make p measurement, each integrated over a period, t, and add them we obtain a result whose signal to noise ratio will be

$$\frac{P_s}{N} = \frac{2v^2 pt}{S} \qquad \ldots(15.12)$$

What matters here is the *Total Measurement Time, pt*, not the choice of each individual period, t. Note also that the choice of the integrator's time constant value, τ, does not affect the signal to noise ratio. In a real measurement situation we should simply choose a τ value which provides a convenient output level after each sample integration period, t. Provided that we avoid voltages which are too large or too small to measure reliably with the voltmeters, etc, we're using the value of τ has no effect on the signal to noise ratio — and hence the accuracy — of the final result.

In practice we're often interested in obtaining a value proportional to the signal voltage (or current) level instead of the power. The integrated output signal voltage increases linearly with pt. However it is the output noise <u>power</u> which increases linearly with time — i.e. the typical output noise voltage increases as $\sqrt{pt}$. Hence the accuracy of a measured voltage will increase in proportion with the square root of the measurement time.

133

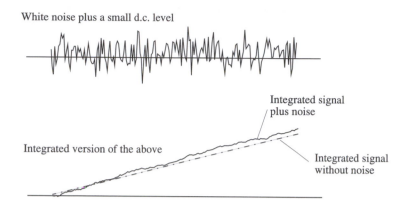

White noise plus a small d.c. level

Integrated signal plus noise

Integrated version of the above

Integrated signal without noise

Figure 15.2   Integrating a steady signal with some superimposed noise.

Figure 15.2 illustrates the effect of integrating an input which consists of a combination of a steady 'd.c.' level plus some white noise. In this case the magnitude of the input d.c. voltage is a quarter of the rms noise voltage. It can be seen that the integrated result allows the steady level to 'grow' linearly with time whilst the effects of noise only change relatively slowly.

The analog integrator is a convenient way to obtain a result averaged over a period of time. In principle we could use a simpler method. For example, we could regularly note down the reading on a voltmeter, then add up all the readings. The result would be a 'piecemeal' value for the level summed or integrated over the period of the readings. Provided that the readings were taken often enough to form a complete record we'd get the same information as if we'd used an analog integrator. No matter what method we use for 'adding up' measurements over the time period the result would be the same. When measuring a signal in the presence of white noise we get a final S/N power ratio which improves linearly with the measurement time (i.e. the signal/noise voltage ratio increases with the square root of the time taken for the measurement).

Although we won't attempt to prove it here, a similar result arises when we look for other signal patterns in the presence of white noise. From an information theory viewpoint a steady (d.c.) level is just one example of a specific signal pattern. Any other pattern can be searched for in the presence of noise. Although we would have to process the signal+noise patterns differently we will discover the same basic result. When the noise is white the final accuracy of a measurement improves with time just as the above example.

134

The above conclusion applies for a white noise spectrum. A different result arises when $1/f$ noise is present. Consider, for example, a case similar to the above but where the noise has a NPSD

$$S\{f\} = \frac{e}{f} \qquad \qquad \text{...(15.13)}$$

The effective noise power observed at the output of the integrator will be

$$N = \int_0^\infty S\{f\} G\{f\} df = \int_0^\infty \left(\frac{e}{f}\right) \frac{\mathrm{Sin}^2\{\pi f T\}}{(\pi f \tau)^2} df \qquad \text{...(15.14)}$$

To see what this integral implies it simplifies things to make a change of variable to $z \equiv \pi f T$. We can then write that

$$N = \frac{e T^2}{\tau^2} \int_0^\infty \frac{\mathrm{Sin}^2\{z\}}{z^3} dz = \frac{e T^2}{\tau^2} \times I \qquad \text{...(15.15)}$$

where I represents the integral in z. Taking the same signal as before the integrated measurement therefore has an effective signal/noise ratio of

$$\frac{P_S}{N} = \left(\frac{v^2}{eI}\right) \qquad \qquad \text{...(15.16)}$$

Note that the integration time does not appear in this expression. This means that we cannot obtain a more accurate result in the presence of $1/f$ noise simply by integrating over a longer period. Worse still, the value of the integral, I, turns out to be infinite!

The integral 'blows up' in this way because we have assumed that the noise power spectral density $\to \infty$ as $f \to 0$. In reality we wouldn't notice the noise components at frequencies $f \ll 1/2T$ as fluctuations. They would look like a fairly steady level during the particular observation time we've used and become indistinguishable from the signal. This simply confirms that we can't get rid of the effects of $1/f$ noise by integrating or adding together lots of measurements.

In practice the noise present in a measurement system will have both white and $1/f$ components. The total noise spectrum can then be represented as a NPSD, S_t, equal to

$$S_t = S + \frac{e}{f} \qquad \qquad \text{...(15.17)}$$

Provided the total measurement time, $pt \ll \frac{S}{e}$, we won't observe any significant effect from the $1/f$ noise as the measurement would be dominated by the white noise. Under these circumstances we can expect to obtain an improvement in measurement accuracy by increasing pt.

135

However, once $pt \geqslant \frac{S}{e}$, the effect of the $1/f$ noise becomes significant and any further increase in pt will produce little or no improvement of the measurement accuracy.

A similar analysis can be carried out for other forms of signal filter and signal summing or integration systems. Although the details will depend upon the choice of system the consequence is much the same. There is a practical limitation — set by the existence of $1/f$ noise — to the improvement in measurement accuracy we can obtain simply by averaging or summing over ever longer periods of time. In order to obtain any further increase in accuracy we must, instead, devise some measurement technique which avoids the effect of $1/f$ noise.

Another serious problem which can arise when making simple, direct measurements is due to the presence of any unwanted *background* signals. Consider as an example the case illustrated in figure 15.1 where we are using a sensor to detect the output from a faint source of light. If we place the source and detector in an ordinary room we find that some of the light striking the detector does not come from the source we wish to measure. Instead it comes from the room lights, or in through the windows of the room. Hence the output we observe from the detector is partly produced by an unwanted 'background'.

One way to deal with this problem is to try and reduce the background level, ideally to nothing. We can, for example, switch off the room lights and place opaque covers over the windows to produce a dark room. Although this means we tend to fall over the furniture it will reduce the unwanted background level. Unfortunately, some background light will remain. This is because the room will be at a temperature above absolute zero. To avoid freezing its inhabitants the room temperature will probably be somewhere around 280 K to 300 K. Hence all of the surfaces in the room will emit some thermal radiation. Unless we totally enclose the detector in a box cooled to absolute zero there will <u>always</u> be some background radiation falling upon it. (And, of course, if we <u>totally</u> enclose it in a box, we can't get the signal onto it!)

Since all sensors and detectors respond to energy or power in one form or another a similar result occurs in every measurement system. We may therefore expect that there will be always be an unwanted background level falling upon any detector. In many cases we can reduce this background until it's low enough to be ignored, however it is impossible to really reduce it to zero.

As an alternative to trying to get rid of the background we can set out to measure it in the absence of the actual signal and then subtract its effect from the final measurement of interest. This approach, called *Background Subtraction*, is widely used to deal with the problems of measuring very small quantities in the presence of an unwanted background.

Summary

You should now know how an *Integrator* works, and how it can be used to improve the S/N ratio of a measurement of a steady signal in the presence of noise. It should be clear that — when the random noise is 'white' — the S/N ratio we can obtain is proportional to the total time devoted to the measurement. That the precise choice of the *Time Constant* of an analog integrator doesn't normally affect the final result. Remember that, since the S/N power ratio improves in proportion with the time, the accuracy of a voltage measurement increases with the square root of the measurement time. You should also now realise that integration doesn't always provide an improvement in the measurement's S/N. In particular, integration does not help us overcome the effects of $1/f$ noise.

Questions

1) Draw a diagram of an *Analog Integrator* and explain how it works. Define the integrator's *Time Constant* in terms of the component values.

2) An analog integrator is constructed using an op-amp, a 100 kΩ resistor, and a 10 µF capacitor. What is the value of the integrator's time constant? What is the value of the integrator's *Power Gain* at 5·25 Hz when used for a 10 second integration? [**1 second. G = 0·003676 or −24·4 dB.**]

3) The integrator described in question 2 is used to determine a steady d.c. level. The input noise spectrum is white and has a NPSD of 10 nV/$\sqrt{\text{Hz}}$. What input d.c. level would be detectable with a 1:1 signal/noise ratio by averaging together 20 measurements, each lasting 5 seconds? (Remember that the expressions in this chapter were simplified by assuming that the effective impedances everywhere all equalled 1 Ohm.) [**0·7 nV.**]

Chapter 16

Phase sensitive detection

The *Phase Sensitive Detection* (PSD) technique is widely used to deal with the problems caused by $1/f$ noise and unwanted background levels. Figure 16.1 represents a typical PSD system, designed to provide a measurement of the signal level produced by a faint light source, S_1. The technique works by arranging for the signal level to be 'switched on and off' in a controlled way. This helps the measurement system distinguish the (now varying) signal from any steady background level.

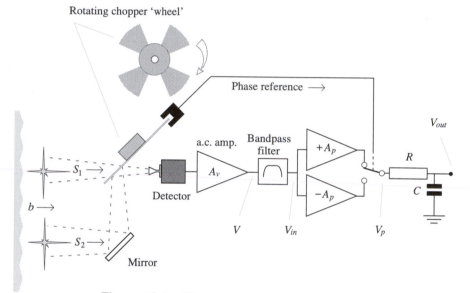

Figure 16.1 Phase sensitive detection system.

Sometimes we can manage to switch a signal source on and off directly by, for example, controlling its power supply. More generally, however, this is not possible. In some cases the source will behave poorly (or fail!) if we keep turning it on and off. Sometimes the source we are interested is a natural one (e.g. a star) which we find rather difficult to control! Therefore most PSD systems employ some form of signal *Modulator* or *Switch* which periodically stops the signal from reaching the detector. When making an optical measurement this modulator can take the form of a *Beam Chopper* which alternately blocks and unblocks the light path

138

between source and detector. Figure 16.1 illustrates one common type of modulator called a *Chopper Wheel*. This is a disc which has a series of *Blades* cut around its periphery. (Sometimes a series of holes are cut around the edge of the wheel to produce a similar effect.) The wheel is placed so that its edge covers the beam and is rotated during the measurement. As the chopper wheel turns, its blades pass between source and detector, alternately blocking and clearing the signal path. If we use a symmetric wheel with n blades, rotating x times per second the source signal reaching the detector will be appear as a fluctuating level, varying periodically with a *Chopping Frequency*, $f = nx$.

The chopper acts as a form of *Frequency Conversion* system. A light power level which was steady or slowly varying, now produces a chopped signal at some higher frequency. The signal power has been converted from one frequency (about d.c. or 0 Hz) to another, f. For the sake of example we can imagine rotating a 16-bladed chopper 20 times a second to produce a chopping frequency of $16 \times 20 = 320$ Hz. This can be amplified using an a.c. amp and passed through a filter arranged to reject signal fluctuations at frequencies below, say, 200 Hz.

Any $1/f$ noise produced in the detector and amplifiers will usually be at frequencies around 100 Hz or less. The filter will stop this low frequency noise from passing through the system. Note we are talking about the frequencies of fluctuations of the signal <u>not</u> the optical frequency of the light itself. The signal we're talking about here corresponds to the voltage/current levels produced by the light <u>power</u> falling upon the detector. To avoid confusion it is customary to refer to the fluctuation frequency produced by modulation as the *Modulation* or *Chopping* frequency. We could now determine the brightness of the light by measuring the size of the signal fluctuations at the chopping frequency emerging from the filter. In this way we can use the PSD system to make a measurement largely unaffected by the $1/f$ noise.

Some further advantages can be obtained by recognising that a periodic alternation has a specific <u>phase</u> as well as a frequency. The PSD technique makes use of this fact to obtain some further improvements over simple direct measurements. The best way to understand the behaviour of the PSD system is to begin by considering what the detector observes when the chopper is blocking the signal path. It's important to realise that the 'source blocked' level is rarely zero. In the simplest optical systems the chopper is painted black or is made of a material which absorbs light. As a result, when a chopper blade blocks the detector's view it sees thermal

139

radiation emitted by the chopper surface. The amount of radiation produced will depend upon the material and its temperature.

Figure 16.1 illustrates the use of a <u>reflecting</u> chopper made of a shiny material. When its blades block the signal beam the detector will see light reflected by the surface of the chopper. In the system shown another mirror is used to direct light from a second source, S_2, onto the detector via reflection from the chopper surface. The system is arranged so that both sources are seen against the common background level, b.

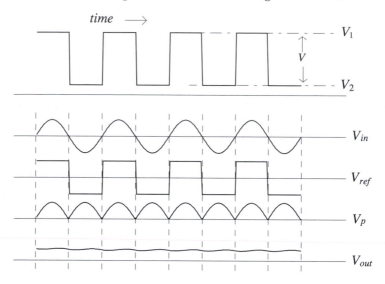

Figure 16.2   Signals in correctly phased PSD system.

If we define S_1, S_2, and b to be the power levels produced by the two sources and the background, a detector whose responsivity is α V/W will produce an output voltage

$$V_1 = \alpha(S_1 + b) \qquad \text{...} (16.1)$$

when the signal path to S_1 is clear, and a voltage

$$V_2 = \alpha(S_2 + b) \qquad \text{...} (16.2)$$

when one of the reflecting blades fills the detector's field of view. The magnitude of the alternating signal, V, output by the detector will therefore be

$$V = \alpha(S_1 - S_2) \qquad \text{...} (16.3)$$

These voltages, and the others in the PSD, are illustrated in figure 16.2.

140

Provided that the background level remains the same no matter which source the detector sees, the magnitude of the alternating voltage, V, is unaffected by the background. The system hence suppresses the effects of any common background level as well as producing an alternating signal. In principle we need not actually employ a second source. If it is omitted ($S_2 = 0$) the magnitude of the output voltage will simply be $V = \alpha S_1$. As will become clear later, however, it can often be a good idea to employ a second source.

In practice the background levels against which the two sources are seen may not be identical. Even when the process producing the background is physically the same in both cases, its level may change with time, making it different during the times when S_1 and S_2 are being observed. Hence we cannot expect to completely suppress any effects due to the background. We can, however, usually arrange to dramatically reduce the influence of background power upon the measurement — provided the system is carefully designed and operated.

In figure 16.1 the output signal from the detector is passed through an a.c. amplifier whose voltage gain is A_v and a bandpass filter which only passes a range of frequencies around the modulation frequency, f. The filtered signal, V_{in}, then passes through an arrangement whose voltage gain can be switched to be either $+A_p$ or $-A_p$. The setting of the switch which selects the sign of this gain is controlled by a *reference* signal, taken from the chopper, which indicates whether the detector can see S_1 or S_2 at any moment. Ideally, the system will be set up so that the signal modulation and the reference signal are *in phase*. This means that the gain will be switched to $+A_p$ while the detector can see S_1 and to $-A_p$ while it can see S_2 and the output switch operates as the chopper blades move in/out of the detector's field of view.

Provided the chopper's teeth and gaps cover an area much larger than the detector's field of view we can assume that the modulated output from the detector will be a square-wave of frequency, f, and peak-to-peak amplitude, V. Now a square-wave of frequency, f, and peak-to-peak amplitude, V_{ptp}, can be regarded as being the sum of a series of sinewaves of the form

$$V = \left(\frac{2V_{ptp}}{\pi}\right).\left(\mathrm{Sin}\left\{2\pi f t\right\} + \frac{1}{3}\mathrm{Sin}\left\{2\pi\left(3f\right)t\right\} + \frac{1}{5}\mathrm{Sin}\left\{2\pi\left(5f\right)t\right\}...\right)$$

$$...(16.4)$$

Hence, if we assume that the filter passes signals at a frequency, f, without

141

loss but totally rejects signals at frequencies of $3f$ and above, the signal, V_{in}, which emerges from the filter will just be

$$V_{in} = \left(\frac{2aA_v}{\pi}\right)(S_1 - S_2)\,\text{Sin}\,\{2\pi f t\} \qquad \ldots (16.5)$$

The reference signal will also vary periodically at the modulation frequency, f. Since both this reference signal and the modulation of the input are produced by the movement of the same device — the chopper — these two signals will have a fixed phase relationship, i.e. the signal and reference are *coherently* related or *phase locked* to one another. In the example illustrated in figure 16.2 we have assumed that the signal and reference are in phase. In this situation the effect of the switched-gain section is just as if the input were full-wave rectified and amplified by A_p, to produce an output

$$V_p\{t\} = \left|\left(\frac{2aA_vA_p}{\pi}\right)(S_1 - S_2)\,\text{Sin}\,\{2\pi f t\}\right| \qquad \ldots (16.6)$$

In the illustrated system this voltage is passed through an RC time constant. Provided $RC \gg \frac{1}{f}$ this time constant circuit will smooth out the half-cycle fluctuations in V_p to produce an output voltage, V_{out}, which will settle at a mean voltage

$$V_{out} = \frac{1}{T}\int_0^T V_p\{t\}\,dt \qquad \ldots (16.7)$$

where $T = \frac{1}{f}$. (This integral gives the right value because each cycle of V_p is the same as all the others. As a result, the average voltage over many cycles is identical with the average voltage over just one cycle.) Putting 16.6 into 16.7 we get the result

$$V_{out} = \left(\frac{2}{\pi}\right)^2 (S_1 - S_2)\,aA_vA_p \qquad \ldots (16.8)$$

V_{out} represents the mean (i.e. time-averaged) voltage level of the output produced by the PSD. The time constant performs the task of allowing this mean level through to the output meter whilst rejecting any voltage fluctuations at frequencies around the modulation frequency, f, or above. When viewed overall, the PSD system converts a steady (or slowly varying) input to an alternating signal at a modulation frequency, f. It then amplifies and filters the signal before reconverting it back into a steady voltage for measurement. The manner in which this is done allows us to largely suppress unwanted background effects and avoid $1/f$ noise generated in the detector and amplifiers.

Provided we know a, A_v, A_p, and S_2, we can determine the light power level, S_1, by measuring V_{out} with a d.c. voltmeter. However, any errors in these quantities will produce a corresponding error in our measurement of S_1. For this reason a better approach is use what is known as a *Nulling measurement* technique. To do this we need to choose a controllable source for S_2. The PSD system illustrated in figure 16.1 compares the light power levels S_1 and S_2 and provides an output signal voltage which varies in proportion with the <u>difference</u> between the two levels. Given a comparison source, S_2, whose output may be varied in a well defined manner we can adjust its output until $V_{out} = 0$. From expression 16.8 this can only arise when $S_1 - S_2 = 0$, no matter what the values of α, A_v, and A_p (assuming, of course, none of them are zero!). Hence, if $V_{out} = 0$ and we know S_2, we can simply say that $S_1 = S_2$ without needing to know any of the amplifier gains or the detector responsivity.

Nulling techniques are very useful when we need to make accurate measurements. They permit us to avoid many of the systematic errors which arise when the behaviour of the amplifiers and sensors are not well known. The technique does of course require us to have a well defined, controllable, reference against which to measure. However in principle <u>all</u> measurements are comparisons — direct or indirect. The nulling measurement simply brings as much as possible of the chain of comparisons within a single system.

Thus far we have assumed that the chopped signal and the reference output share the same phase. This may not always be the case. Consider the situation when, for some reason, the signal and reference waveforms differ in phase by an amount ϕ. If we define the time, t, such that $t = 0$ corresponds to a moment when the chopper moves out of the detector's field of view then we can show that the output from the switched gain circuit, V_p, will be

$$V_p\{t\} = \left(\frac{2\alpha A_v A_p}{\pi}\right)(S_1 - S_2)\operatorname{Sin}\{\Theta - \phi\} \text{ when } \operatorname{Sin}\{\Theta\} > 0$$

$$\text{or} = -\left(\frac{2\alpha A_v A_p}{\pi}\right)(S_1 - S_2)\operatorname{Sin}\{\Theta - \phi\} \text{ when } \operatorname{Sin}\{\Theta\} \leqslant 0$$

$$\text{...} (16.9)$$

where $\Theta = 2\pi f t$.

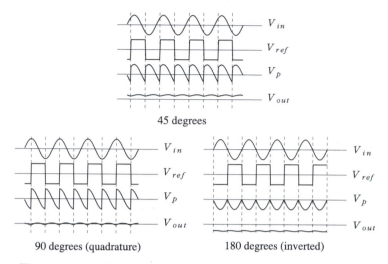

Figure 16.3   Effect of various phase errors on PSD signals.

Using expression 16.7 we can therefore expect that the smoothed voltage, V_{out}, this presents to the voltmeter will be

$$V_{out} = \left(\frac{2}{\pi}\right)^2 \alpha A_v A_p (S_1 - S_2) \, \text{Cos}\{\phi\} \qquad \ldots (16.10)$$

i.e. we find that the magnitude of the smoothed output voltage will vary in proportion with the cosine of the *phase error*, ϕ. The effects of various phase error values are illustrated in figure 16.3. This result has two implications. Firstly, it is clearly important to ensure that the PSD is 'phased up' correctly — i.e. we should adjust the system to ensure that the wanted signal and the reference share the same phase — otherwise the magnitude of the signal output will be reduced by $\text{Cos}\{\phi\}$. The second implication concerns the system's ability to reject noise or any other signals at frequencies which differ from f.

The noise produced in the detector, amplifiers, etc, can be regarded as a spectrum of components at various frequencies. If we were to observe the noise voltage generated during some specific period of time it could then (from the sampling theorem arguments) be described as a spectrum of the general form

$$V_n = \sum_{i=1}^{N} \Delta V_i \, \text{Cos}\{2\pi f_i t + \Phi_i\} \qquad \ldots (16.11)$$

where the ΔV_i and Φ_i values vary unpredictably from one noise

144

observation period to another. From the statistical properties of noise we can expect the average value of ΔV^2 to depend upon the mean noise power level. The phases can take any values. Since there is a bandpass filter in the system we need only worry about noise components at frequencies similar to the signal chopping frequency, f. We can therefore consider two situations. Firstly, consider the noise component <u>at</u> the chopping frequency. The above description of the observed noise can be re-written as

$$V_n = \sum_{i=1}^{N} A_i \operatorname{Cos}\{2\pi f_i t\} + B_i \operatorname{Sin}\{2\pi f_i t\} \qquad \dots (16.12)$$

where

$$A_i = \Delta V_i \operatorname{Cos}\{\Phi_i\} \quad ; \quad B_i = \Delta V_i \operatorname{Sin}\{\Phi_i\} \qquad \dots (16.13)$$

i.e

$$V_n^2 = A_i^2 + B_i^2 \qquad \dots (16.14)$$

when considering noise at the signal frequency $f = f_i$. From the behaviour of phase sensitive detection only the in-phase portion of the noise, $A_i \operatorname{Cos}\{2\pi f_i t\}$, will have any effect upon the output. The quadrature portion, $B_i \operatorname{Sin}\{2\pi f_i t\}$, will produce no output. Since the noise phase varies at random we find that, on average, $A_i^2 \approx B_i^2$. This means that, on average, $A_i^2 = \Delta V_i^2 / 2$; i.e. only <u>half</u> the input noise power has an effect upon the output. As a result, the PSD has the effect of rejecting that half of the input noise at the signal frequency which is *In Quadrature* with the signal. This means that the system gives a better signal/noise ratio than we would've obtained if we'd simply measured the size of the chopped a.c. signal with an a.c. voltmeter.

Consider now a noise component whose frequency, f_i, differs from the chopping frequency by an amount, δf — i.e. the noise component can be written as

$$\Delta V_i \operatorname{Cos}\{2\pi (f + \delta f) t + \Phi_i\} \qquad \dots (16.15)$$

Looking up trig identities in a suitable maths book we can find this is equivalent to

$$\Delta V_i [\operatorname{Cos}\{2\pi f t\} \operatorname{Cos}\{2\pi \delta f t + \Phi_i\} + \operatorname{Sin}\{2\pi f t\} \operatorname{Sin}\{2\pi \delta f t + \Phi_i\}]$$

$$\dots (16.16)$$

Because of the action of the PSD only the Cos part of this has an influence upon the output. In effect, it is equivalent to an input

$$\Delta V_i' \operatorname{Cos}\{2\pi f t\} \qquad \dots (16.17)$$

145

where

$$\Delta V_i' = \Delta V_i \, \text{Cos}\{2\pi\delta f t + \Phi_i\} \qquad \ldots (16.18)$$

i.e. the noise component produces an output which varies sinusoidally at the *Beat Frequency* or *Difference Frequency*, $\delta f = |f_i - f|$. This effect is illustrated in figure 16.4. Here the signal/reference and the noise component 'beat in and out of phase' with each other and produce a smoothed output level which varies roughly sinusoidally. For example, if we are using a chopping frequency of $f = 1000$ Hz, noise at $f_i = 1001$ Hz will cause the output to vary sinusoidally at 1 Hz. Note that noise at 999 Hz will <u>also</u> produce output at 1 Hz when the chopping frequency is 1000 Hz.

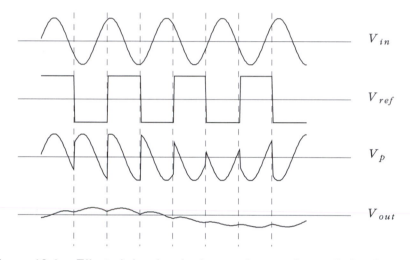

Figure 16.4 Effect of signal and reference frequencies not being the same.

Now the output resistor-capacitor *time constant* acts as a low-pass filter. It will only pass signal or noise fluctuations in the frequency range from d.c. (0 Hz) up to $\frac{1}{2\pi\tau}$ Hz, where $\tau = RC$ is the time constant value of the filter. For example, a 1 second time constant (perhaps made with a 10,000 Ω resistor and a 100 μF capacitor) will only pass unattenuated frequencies below 0.159 Hz. This means that the output level will only be affected by signal noise in the frequency range 0.159 Hz either side of the chosen chopping frequency. In effect, the output time constant acts just like a very narrow bandpass filter to block noise at frequencies which differ from the signal we're interested in. In theory the same result could be obtained using a very narrow-band filter. In practice using the output time constant has two advantages. Firstly, it is possible to build time constants (or *integrators* which produce a similar result) with time constants of many

146

seconds. To achieve the same results when using a 1 kHz chopping frequency we would have to build a bandpass filter with a bandwidth of much less than 1 Hz at 1 kHz. Although not impossible, this would be much harder to make.

Secondly, the output time constant does not mind if the chopping frequency should alter slightly for any reason. The speed of the chopper might slowly drift as its motor warms up. If we used a narrow bandpass filter we'd have to ensure that the chopping frequency doesn't drift so far as to shift the signal frequency outside the filter's passband. Otherwise the signal will be lost! Since the PSD switch is controlled by the chopper, any change in chopping frequency will be cause the switching action to alter so as to take the change into account. For these reasons, plus the PSD's ability to reject quadrature noise, the majority of the noise filtering action of a PSD is performed by the output time constant or integrator.

This being the case it's sensible to wonder why we should bother to include a bandpass filter at all. There are two reasons for including it. Firstly, sometimes the total input broadband noise power may be much larger than the signal power. Unless we filter away some of this noise it will limit the amount of amplifier gain we can use because, otherwise, it will *saturate* or clip the amplifiers. Secondly, if we go through the same analysis as above but with a frequency which is an odd harmonic ($3f$, $5f$, etc) of the chopping frequency, we find that the switching action causes these to produce output at frequencies low enough to pass through the output time constant. Hence noise at these frequencies won't be blocked by the output filter. The bandpass filter stops detector and amplifier noise at these frequencies from reaching the output.

PSDs are used in many forms for measurement and information processing tasks. They are an example of a *Heterodyne* system. Similar techniques are used in radios, TVs, and radars. Radioastronomers use a technique called *Dicke Switching* and optical astronomers use *Sky Chopping* or *Telescope Nodding* to achieve the same results. The method is useful whenever we wish to alter the signal frequency to avoid noise, make the information more easily handleable, or suppress background effects. The special technique of *Nulling* is also one of the most reliable ways to make very accurate measurements.

Summary

You should now know how *Phase Sensitive Detection* (PSD) systems work. That they can be used to avoid $1/f$ noise in detectors and amplifiers and can be used to subtract the effects of a steady background level. You should also now understand that the PSD is an example of a *Heterodyne* technique which uses *Frequency Conversion*. You should also know that a *Nulling* measurement technique is useful because it means we don't have to know exactly the sensitivity or gain of our detectors and amplifiers when making accurate measurements.

Chapter 17

Synchronous integration

17.1 'Boxcar' detection systems

Phase sensitive detection systems are ideally suited to dealing with signals which have a steady, or relatively slowly varying, level. In many situations, however, we need to measure the details of a signal which varies quite swiftly in a complex manner. The signal may also not last very long. In order to measure brief, rapidly changing signals a different approach is required. *Synchronous Integration* is a technique which allows measurements to be made on complex signal patterns which have powers well below the general detector or amplifier noise level. The technique can be employed in various ways provided two basic requirements are obeyed. Firstly, the signal must be <u>repeatable</u> so we can produce a series of nominally identical pulses or *Signal Cycles*. Secondly, we must obtain an extra *Trigger* signal — similar to the phase reference signal required for a PSD — which can be used to tell the measurement system when each signal cycle begins. Although it's usually convenient to arrange for signal cycles to occur with a steady repetition rate, this isn't absolutely necessary provided we know when each cycle starts.

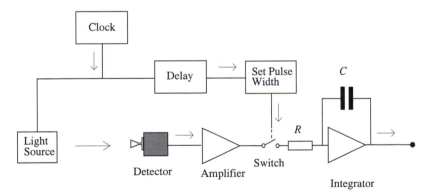

Figure 17.1 Analog synchronous integration (boxcar) system.

These requirements are often satisfied by using some form of *clock* which regularly initiates the signal and provides the trigger information. Alternatively, the signal generating process may, in itself, provide some

149

information telling us when each signal cycle begins. For the sake of illustration we can concentrate upon a situation where we wish to measure how the output light intensity of a pulsed laser varies with time during each output signal pulse. The techniques described in this chapter can, however, be applied to measure the shape of any repetitive signal pattern.

Some electrical gas discharge lasers can be arranged to produce a series of light pulses when connected, via a suitable circuit, to a steady power supply. Each burst of light output is accompanied by an abrupt drop in the voltage across the gas tube. Under these circumstances we could use the sudden fall in voltage to trigger the measurement process. More generally, however, we will have to provide some kind of clock signal to initiate light output. Figure 17.1 illustrates a typical system designed to measure how the output intensity of a pulsed laser varies with time. In this case we have arranged for the system to be controlled by a clock which both 'fires' the laser and triggers the measurements.

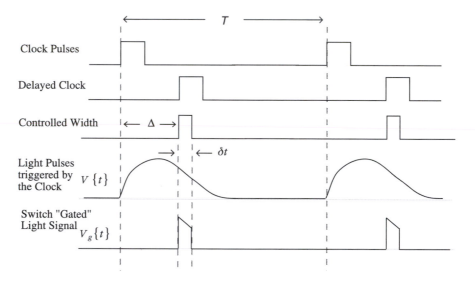

Figure 17.2 Control and data waveforms in 'boxcar' integrator.

For the sake of simplicity we can assume that the clock which starts each cycle of light output has a period, T. This means that the resulting signal cycles will occur at the rate, $1/T$. Each clock pulse immediately starts a signal cycle. The clock also controls the operation of a switch which can connect the amplified signal to an analog integrator. The switch is only closed for a brief *Sampling Interval*, δt, which begins after a time delay, Δ,

150

following the appearance of each clock pulse.

Synchronous integration works on the basis that all the signal cycles are similar to one another. We can then define the shape of each individual pulse in terms of the same function, $v\{t\}$, where t represents the time from the beginning of each signal cycle. Figure 17.2 illustrates a typical set of pulse and signal patterns we might see in a working system of this kind. The output voltage from the detector is amplified to produce a signal voltage, $V\{t\}$, which is presented to the switch. Since the switch is only connected for a brief period, δt, after a delay, Δ, following the start of each clock pulse, the signal presented to the integrator looks like the waveform, $V_g\{t\}$, shown in figure 17.2. This can be defined as

$$V_g\{t\} \equiv V\{t\} \quad \text{when } \Delta \leqslant t \leqslant \Delta + \delta t$$

$$\text{otherwise } V_g\{t\} \equiv 0 \qquad \qquad \dots (17.1)$$

We can now start with the integrator (capacitor) voltage set to zero and allow the system to operate for n signal cycles. In the absence of any noise this will produce an output voltage

$$V_o\{\Delta, \delta t\} = nK \int_0^T V_g\{t\} \, dt = nK \int_\Delta^{\Delta + \delta t} V\{t\} \, dt \quad \dots (17.2)$$

where

$$K = \frac{-1}{RC} \qquad \qquad \dots (17.3)$$

and R and C are the values of the resistor and capacitor used in the analog integrator. The minus sign is present because an analog integrator normally reverses the sign of the signal (see Chapter 15). Provided δt is sufficiently small, the signal level will not change a great deal between the times, Δ and $\Delta + \delta t$, and we can approximate the above integral to say that

$$V_o\{\Delta\} = nKV\{t\}\delta t \qquad \qquad \dots (17.4)$$

i.e., $V_o\{\Delta\}$, is proportional to the signal voltage, $V\{t\}$, which arises at a time, $t = \Delta$, following the start of each pulse. The output is also proportional to $nK\delta t$, hence we may increase the magnitude of $V_o\{t\}$ by operating the system for more clock cycles, increasing the value of n. In effect the system adds up the contributions from a series of pulses to magnify the output signal level.

In practice, the required signal will always be accompanied by some unwanted noise voltage, $e\{t\}$, which — being random — will differ from one pulse to another. This will contribute an unpredictable amount

151

$$E_o = K \sum_{i=1}^{n} \int_{\Delta}^{\Delta + \delta t} e\{iT + t\} \, dt \qquad \qquad ...(17.5)$$

to the integrated output voltage, where $e\{iT + t\}$ represents the noise voltage during the ith pulse at a time, t, from its start.

Unlike the signal, these noise voltages which occur during each cycle are not all identical. As the noise is random in nature we can't say what value this error voltage will have when we make a particular measurement. As with all random quantities we can only predict the average, typical, or likely properties of the noise. Taking the simplest example of a 'white' noise input spectrum whose noise power spectral density is S. We can use the arguments presented in section 15.2 to say that the mean noise power added to a single integration will be $N_i = K^2 S \delta t / 2$. (This result comes from considering expression 15.9 and recognising that, in this case, the integration constant $K^2 \equiv 1 / \tau^2$.) This means that the voltage produced by each individual sample integration will typically differ from the next by a rms amount

$$\varepsilon_n = \sqrt{N_i} = K \sqrt{\frac{S \delta t}{2}} \qquad \qquad ...(17.6)$$

The noise power spectrum of a real white noise source can never extend over an infinite frequency range. (If it did, its total power would be infinite!) For a practical noise source we can therefore say that the input total noise power will be $N_{in} = S B_n$, where B_n represents the *Noise Equivalent Bandwidth* of the input noise spectrum. Here we can assume that this means that the noise covers the frequency range from around d.c. (0 Hz) up to a maximum frequency equal to B_n. The input will therefore exhibit an input noise voltage level equivalent to an rms voltage of $e_n = \sqrt{S B_n}$.

Combining these expressions we can therefore say that the input and output rms noise voltage levels will be such that

$$\varepsilon_n = K e_n \sqrt{\frac{\delta t}{2 B_n}} \qquad \qquad ...(17.7)$$

This expression links the rms noise level, ε_n, at the integrator's output to the input level, e_n. We can now use this expression to determine the accuracy of a measurement using the synchronous integrator, although it is worth remembering that, in general, the precise relationship between ε_n and e_n depends upon the details of the input noise spectrum. A more detailed analysis would show that expression 17.7 is only strictly true for a noise spectrum which has a uniform noise power spectral density over a frequency range, f_{min} to f_{max} where $f_{min} \lll \frac{1}{2\delta t}$ and $f_{max} \ggg \frac{1}{2\delta t}$.

As the actual noise level varies randomly from one measurement to another we can say that typical measured levels after n signal cycles will be

$$V_o'\{\Delta\} = nKV\{\Delta\}\delta t \pm \varepsilon_n \sqrt{n} \qquad \qquad ...(17.8)$$

The unpredictability of the noise means we can't predict a precise value for V. Instead, expression 17.8 indicates the most probably result, plus or minus the probable range of uncertainty. Here the prime indicates a typical measured value which may not exactly equal the result we might predict using expression 17.4. Combining expressions 17.4, 17.7 and 17.8 we can obtain

$$V_o'\{\Delta\} - V_o\{\Delta\} = \pm Ke_n \sqrt{\frac{n\delta t}{2B_n}} \qquad \qquad ...(17.9)$$

In effect this shows the probable difference between the values we would measure with and without random noise.

From expression 17.4 we could expect — in the absence of any random noise — to find the input signal voltage level, $V\{t\}$ at a time $t = \Delta$ from the expression

$$V\{t\} = \frac{V_o\{\Delta\}}{nK\delta t} \qquad \qquad ...(17.10)$$

unfortunately, the inevitable presence of some noise means that a typical measurement leads to the actual result

$$V'\{t\} = \frac{V_o'\{\Delta\}}{nK\delta t} \qquad \qquad ...(17.11)$$

Combining expressions 17.9–17.11 we can say that our measurement of the input voltage at any time will be

$$V'\{t\} = V\{t\} \pm e_n \sqrt{\frac{1}{2nB_n\delta t}} \qquad \qquad ...(17.12)$$

From 17.12 we see that the accuracy of measurements of the input signal level will tend to improve as we increase the number of signal cycles we integrate over. Two points about this result are worth noting. Firstly, both the total input noise level __and__ the frequency range it covers affect the accuracy of the measurement. This can be understood by imagining a situation where a given fixed total input noise power is 'stretched out' to cover a wider frequency range. The effect of such a change would be to move some of the noise power up to higher frequencies which find it

153

more difficult to pass through an integrator. Hence the fraction of the noise which influences the output will fall if B_n is increased while e_n is kept constant. Secondly, the above result indicates the relative sizes of the measured signal and noise voltages. When considering the performance of a signal processing system in terms of S/N ratios we normally consider a power ratio. Since the voltage accuracy obtained above varies as $\sqrt{\delta t\, n}$ we can expect the output S/N (power) ratio provided by a synchronous integration system to improve with $\delta t\, n$ — i.e. in proportion with the number of signal cycles integrated.

In order to measure the overall shape of the signal waveform — and hence the way the laser intensity varies with time — we can now proceed as follows:

Firstly, set Δ to a particular value, zero the integrator voltage, and perform an integration over n clock cycles. Note the integrator output level, increment Δ by an amount, δt, and rezero the integrator. Integrate again for n cycles, and note the new output level. Repeat this process until a series of $V_o'\{\Delta\}$ values have been gathered which cover the whole of the signal cycle. Then use expression 17.11 for a set of times, $t = \Delta$, to determine the shape of the input signal with an accuracy which can be estimated using expression 17.12.

This form of measurement system is called a synchronous integrator because we perform integrations on samples which are *synchronised* with the signal cycles. Many of the earliest system employed an output time-constant instead of an integrator. The time delay, Δ, was then slowly swept continuously over the range 0 to T and the smoothed output displayed on an oscilloscope or drawn on a plotter. These systems came to be called 'boxcar' integrators because the switch control pulse looked on an oscilloscope like an American railroad waggon running along a track.

Synchronous integration systems are very effective at recovering information about weak pulses when the noise level is quite high. As usual, however, there is a price to be paid for this improvement in the measured S/N ratio. The total measurement for any particular delay, Δ, takes a time nT since we have to add up the effects of n clock cycles. Hence when we improve the S/N ratio by increasing n, the measurement takes longer. A drawback of the method considered so far is that most of the time the output integrator is disconnected from the input! Only that fraction, $\delta t/T$, of the pulses which occur while the switch is closed contributes to the measurement result. As a consequence, to measure all the details of the pulse shape we have to repeat the measurement process up to $T/\delta t$ times for each Δ value. Hence the time required to measure the whole signal shape will be $nT^2/\delta t$. If n is large and δt small, this can turn out to be quite a long while!

To improve the S/N ratio without increasing the total measurement time we could chose to increase, δt, the duration of each sample. Unfortunately we can't expect to observe any signal fluctuations which take place in a time-scale less than δt because they will be smoothed away by the integrator. When using a synchronous integrator we can only clearly observe details of the pulse shape which persist for a time $\geqslant \delta t$. We can therefore reduce the total measurement time by increasing δt, but this may mean that we can no longer see all of the fine details of the signal. Any real signal will only contain frequency components up to some finite maximum frequency, f_{max}. From the arguments outlined in chapter 2 (section 4) we can expect that we will only be able to see all the details of the signal when

$$\delta t \leqslant \frac{1}{2f_{max}} \qquad \qquad ...(17.13)$$

In practice, therefore, $\frac{1}{2f_{max}}$ usually represents the optimum choice for δt. A smaller value increases the required measurement time, a larger value prevents us from observing all the details of the signal.

17.2 Multiplexed and digital systems

The system we have considered so far isn't a very efficient one since, in general, most of the signal power was ignored because it arrived when the switch was open. This problem can be dealt with by employing a *Multiplexed* arrangement.

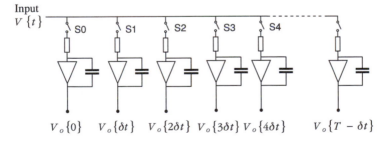

Figure 17.3 Multiplexed array of synchronous integrators.

155

Figure 17.3 illustrates a multiplexed analog synchronous integration system. This works in a similar way to the one we have already considered, but it contains a 'bank' of similar switches and integrators. In this system the first switch, S0, is closed during the periods when $0 < t \leqslant \delta t$, S1 when $\delta t < t \leqslant 2\delta t$, S2 when $2\delta t < t \leqslant 3\delta t$, etc. By using an array of M such switches and integrators, where $M = T / \delta t$, we can arrange that at any time during each pulse one or another of the switches will be closed and the signal is being integrated somewhere. At a time, t, during each pulse the jth switch will be closed, where j can be defined as the integer value (i.e. the 'switch number') such that $j\delta t \leqslant t < (j + 1)\delta t$. Each switch/integrator provides a separate sampling and integration *channel.*

The simple system we considered earlier had just one channel and could only look at a small part of the signal pulse at a time. The fully multiplexed version has $T / \delta t$ channels and covers the whole signal cycle. The system essentially produces a series of integrated output voltages, $V_o\{0\}$, $V_o\{\delta t\}$, etc, and gathers information about all the pulse features 'in parallel'. The advantage of this arrangement is that all of the information from each signal cycle is recorded by the bank of integrators. No signal information is wasted. As a result, the multiplexed system is much more efficient at collecting information than the single-channel version. Using this arrangement we don't have to keep repeating the integration process as Δ is varied.

Although multiplexing means that measurements can be made more quickly and efficiently, wholly analog systems of this type are now rarely used. This is partly because it can be difficult (and expensive) to arrange for a large number of nominally identical switches and analog integrators, but it is also because digital information processing techniques have advanced rapidly over the last few decades. Modern synchronous integration systems often use digital techniques to obtain, relatively cheaply, a level of usefulness it would be difficult to match using analog methods. As usual in information processing we can build various types of digital and analog systems to perform a given function. The system shown in figure 17.4 makes use of a circuit known as a *voltage to frequency convertor* (VFC) to implement a digital synchronous integration system. This is a device which produces an output square wave (or stream of pulses) whose frequency or 'pulse rate' is proportional to the input voltage. At any time, t, we can therefore expect the VFC to be producing pulses at a rate

$$f\{t\} = k_f V\{t\} \qquad \qquad \text{...(17.14)}$$

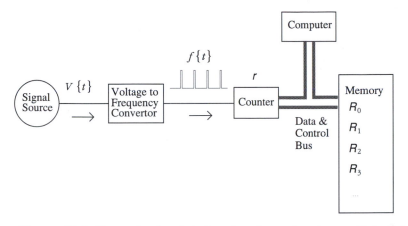

Figure 17.4 Example of a digital system for performing multiplexed synchronous integration of a repetitive waveform.

where k_f is a coefficient whose value depends upon the details of the VFC circuit being used. The operation of this system depends upon how we have programmed the computer. At the start of a measurement the computer should 'clear' (i.e. set to zero) the numbers stored in the parts of its memory which it will use for data collection. The computer then waits until it receives a trigger from the clock which is initiating the pulses to be measured (this can, if we wish, be the computer's own internal clock). The computer then proceeds as follows:

Firstly, the counter reading is zeroed. It is then allowed to count pulses coming from the VFC for a time, δt, and the resulting number, r_0, is added into a memory location at some *address*, A_0. The counter is then re-zeroed, allowed to count for another period, δt, and the new result, r_1, added into a memory location, A_1. This process is repeated over and over again until the whole signal cycle time, T, has elapsed. After one signal cycle the system will have stored a set of binary numbers, r_0, r_1, etc, in its memory. Each number will be approximately equal to

$$r_j = k_f \int_{j\delta t}^{(j+1)\delta t} V\{t\} \, dt \qquad \ldots (17.15)$$

i.e. each number is proportional to the input voltage integrated over a short period of time. We can now repeat this process n times to obtain a stored set of numbers, R_0, R_1, ... , which, in the absence of any noise, will approximate to

157

$$R_j = N r_j = n k_f \int_{j\delta t}^{(j+1)\delta t} V\{t\}\, dt \qquad \dots (17.16)$$

In effect, these stored numbers are proportional to the integrated signal voltages at various times from the start of each signal cycle. They contain the same information about the signal pattern as we could have collected with an analog synchronous integration system. As with the analog system, if we arrange for δt to be small enough we can approximate the above integral to

$$R_j = n k_f \delta t\, V\{t_j\} \qquad \dots (17.17)$$

where $t_j = j\,\delta t$. We can therefore use the collected R_j values to determine the signal voltage at various times during each signal cycle.

The counted values are a digital equivalent of the voltages collected at the output of a bank of analog integrators. Equation 17.17 is the 'digital equivalent' of expression 17.4 for an analog system. Each count is proportional to the input at the appropriate moment, $V\{t_j\}$.

This digital approach has a number of advantages over the analog technique. One particular advantage of the digital approach is that it is relatively easy to buy and use a large amount of computer memory. For example, we can imagine buying and using a single digital memory chip capable of holding 128 *kilobytes* of information. If we allocate 16 bits (i.e. two 8-bit bytes) to hold each R_j we can store a set of values which represent integrated level measurements of the input signal shape at $64 \times 1024 = 65,536$ moments during each pulse. As a result, one cheap digital memory chip can replace over 65 thousand separate analog integrators!

Summary

You should now understand how *Synchronous Integration* allows us to recover the details of a weak, transient phenomenon by adding together the information from a synchronised sequence of similar transient events. That a *Multiplexed* system allows us to avoid the signal information losses we get with a 'single integrator' system which tends to ignore most of the signal most of the time. That we can build either analog or digital systems to perform synchronous integration. You should now also see that the combination of a *Voltage to Frequency Convertor* and a *Counter* act as a form of integrator.

Chapter 18

Data compression

Up until now we've considered systems which always try to preserve all the information content of a message. For example, the CD digital system attempts to digitally encode information in a way which accurately represents all the nuances of any input audio waveforms that fits within a 20 kHz passband and a 95 dB dynamic range. To do this for two channels (stereo) we record and replay $2 \times 16 \times 44{,}100 = 1{,}411{,}200$ bits per second. However, as we discovered in an earlier chapter, some messages aren't very surprising (or interesting) and therefore don't contain much 'real' information. This raises two questions:

> i) Can we re-code a signal into a form which can be sent or stored using fewer bits or bytes without losing any real information?
>
> ii) Do we have to carry all the details of a signal — or can we discard details which are trivial, or 'uninformative'?

The answers to these questions are important because, if we can reduce the amount of bits required, we can send or store useful messages with equipment which has a lower capacity (i.e. cheaper!). The term *Data Compression* has come into use to indicate techniques which attempt to 'Stuff a quart into a pint pot'. Unfortunately, this term is used for a variety of methods which actually divide into two distinct classes. Genuine data compression methods attempt to cut down the amount of bits required without losing any actual information. Other techniques, which I'll call *Data Reduction* or *Data Thinning*, seek out and discard information which they judge 'unimportant'. Data thinning does throw away some real information, but if it works well the lost information isn't missed! In this chapter we'll look at *Lossless* data compression. We'll consider data thinning in the next chapter.

18.1 Run-length encoding

If you use a computer very often you'll eventually encounter the problem of running out of 'space' on the discs you're using to store files of information. Given the cash, this can be solved by buying another box of floppies or getting a larger hard disc. A popular alternative is to use some kind of 'file compression' technique. These often let you squeeze about twice as much onto a disc before it fills up. Various techniques are used

159

for this and some work better than others. Here we'll look at a simple example based on the way computers often store pictures and use it to see the features all true data compression techniques share.

Broadly speaking, computers can store information about images (pictures) either as a set of *Objects* or as a *Pixel Map*. Although object-based techniques tend to give better results we'll look at pixel methods since they provide a clearer example of how data compression can work. Pixel mapping divides the image up into an array of rectangular or square 'picture elements'. The colour of each pixel is stored as a number in the computer's memory. The amount of information (details, range of colours) the picture can contain is then determined by the number of pixels in the image and the range of numbers we can store to indicate the colour of each pixel.

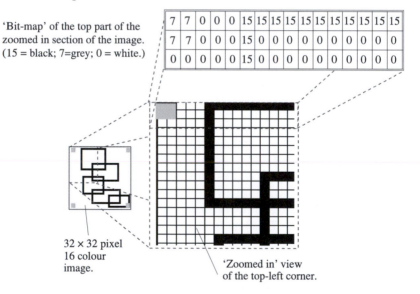

'Bit-map' of the top part of the zoomed in section of the image. (15 = black; 7=grey; 0 = white.)

7	7	0	0	0	15	15	15	15	15	15	15	15	15	15	15
7	7	0	0	0	15	0	0	0	0	0	0	0	0	0	0
0	0	0	0	0	15	0	0	0	0	0	0	0	0	0	0

32 × 32 pixel 16 colour image.

'Zoomed in' view of the top-left corner.

Figure 18.1   Example of a 'bit-mapped' image.

Figure 18.1 shows an example of a bit-mapped image. For the sake of simplicity, we've limited the range of possible colours to just 16 and chosen an image which is only 32 × 32 pixels in size. (Note that, for computers, 'colours' include black, white, and shades of grey.) In fact, modern computers can usually cope with '24-bit' pixel maps. These represent the colour of each pixel as 3 × 8-bit numbers — one 8-bit byte each for the red, blue, and green levels. However, here we're just using <u>four</u> bits per pixel and only using the values to indicate the 'greyscale

160

level' (i.e. how dark the pixel is). This makes the explanation easier, but the following arguments also apply to full-colour pixel-map systems.

In the image shown, the darkness of each black/grey/white pixel is stored as a number in the range 0 – 15 or %0000 to %1111 in binary. 0 means white, 15 black, and 7 a middling grey. (Note we're using a leading '%' to indicate a binary number.) We only need half an 8-bit byte to store the information about each pixel. Since there are 32×32 pixels, each requiring 4 bits, we need a total of $32 \times 32 \times 4 = 4096$ bits or 512 bytes to specify all the details of the image as a bit-map. There are various ways we could record this information on a floppy disc or transmit is over a digital signal link. For example, we can start in the top-left corner and group the pixel values together in pairs to get a string of 8-bit bytes.

The first pair (furthest top-left) of pixel colour numbers are 7 and 7 (%0111 and %0111). Grouping these bits together we get %01110111 = 119. Moving to the left, the next pair of colour numbers are 0 and 0 (%0000 and %0000) which group to produce 0. The next left pair are 0 and 15 (%0000 and %1111) which group to %00001111 = 15. The next pair are 15 and 15 (%1111 and %1111) which group to %11111111 = 255. And so on... having finished the top line we can repeat the left-to-right grouping process line by line down the image. We can therefore store, record or transmit information about the picture's pattern as the series of bytes; 119, 0, 15, 255, etc... The number of bits or bytes required is determined by the 'size' of the picture. To represent <u>any</u> 32×32 pixel, 16 colour pattern we use 512×8-bit bytes. This sort of coding is called *Fixed Length* because the number of bits/bytes required is fixed by the number of pixels and doesn't depend upon the actual picture pattern. A blank (boring) screen — all '0's or all '15's — requires as many bytes as a pretty (interesting) picture.

So, can we store or send all the picture information using fewer bits/bytes? The answer is, yes, <u>sometimes</u> we <u>can</u> by using a different way to code or represent the information. The technique we'll use here is called *Run-Length Encoding*. This is based upon only storing information about where in the picture the colour (darkess/brighness in this case) level <u>changes</u>. To illustrate this process let's assume that the small array of numbers in figure 18.1 is the whole image we want to store or communicate. As shown, this pattern is just 16 pixels wide and 3 pixels high. So we would require $16 \times 3 \times 4 = 192$ bits (24 bytes) to store or send it using the fixed length method described above.

Looking at the example in 18.1 the 16 × 3 bit-map corresponds to the series of byte values: (first line) 119, 0, 15, 255, 255, 255, 255, 255; (second line) 119, 0, 15, 0, 0, 0, 0, 0; (third line) 0, 0, 15, 0, 0, 0, 0, 0. To run-length encode this information we proceed as follows: Begin with the 'first' value (the top-left byte). We note its value — 119 — and then note how many successive bytes have this same value — in this case just 1. We then note the next byte value — 0 — and note how many times it appears in succession — again 1 in this case. We keep repeating this process and generate the values: 119, 1 time; followed 0, 1 time; 255, 5 times; 119, 1 time; 0, 1 time; 15, 1 time; 0, 7 times; 15, 1 time; and 0, 5 times. (Note that we ignored the locations of the line ends/starts and just treated the numbers as one long sequence of bytes.)

This process has produced the series of byte values: 119, 1, 0, 1, 255, 5, 119, 1, 0, 1, 15, 1, 0, 7, 15, 1, 0, 5. Note that this list is only 18 bytes long, yet — provided we know the details of the run-length encoding process — we can use it to reconstruct <u>all</u> of the original 24 bytes of picture information. We have managed to squeeze 24 bytes of information into just 18. At first sight this process seems suspiciously like magic. But it works! We often find that this type of encoding can reduce the number of bits or bytes required to store all the details of a picture. Similar (but more complex) methods can be used to reduce the number of bytes needed to store various sets of information.

The reason this magic trick is possible can be understood by considering two 'extreme' examples of pictures. First consider an image which just consisted of a single black pixel in the middle of a 32 × 32 16-colour bit-map. Recorded as a series of pixel-pair bytes this would be something like: 0, 0, 0, … (about 255 times), 15, then 0, 0, 0, (about 255 times), i.e. 512 bytes consisting of a string of zeros with just one 15 somewhere in the middle. Run-length encoded the same picture information would be something like: 0, 254 times; 15, 1 time; 0, 255 times — just six bytes!

Now consider a picture where every pixel is a different colour to its neighbours. As a plain fixed length series this might be 512 bytes with a pattern something like; 127, 203, 96, 229, etc… Run-length encoded it becomes; 127, 1 time; 203, 1 time; 96, 1 time; 229, 1 time; etc. The run-length sequence now contains <u>1024</u> bytes — twice as many as the plain set of values! This is because we had to dutifully include an extra byte after every pixel value to confirm that the value only appears once before a different one occurs. We can make two general points from these examples:

- The encoding 'doesn't always work' — i.e. it sometimes produces an output series of values which is <u>longer</u> than the fixed length original.
- The degree of compression (or unwanted extension!) depends upon the details of the picture.

Before run-length encoding <u>any</u> 32 × 32 pixel × 16 colour image would be 512 bytes long. After encoding, some images are shorter, some are longer. We've turned a fixed length input *string* of symbols or bytes to a variable length output string. In fact, if we were to repeat the encoding process for every possible, randomly chosen, picture pattern we would discover that <u>on average</u> the compression technique produces an output which is about the same length as the fixed length input. For a randomly chosen message the process shuffles the values but leaves us with the same number of bits to store or communicate. However, most real picture patterns aren't random! Provided we only use the run-length method for pictures which contain many regions where the colour is the same from pixel to pixel the result will be a reduction in the number of bytes. The image shown in 18.1 has a number of large areas of white, so it compresses reasonably well using the run-length method.

18.2 Huffman coding

Although I won't attempt to prove it here, data compression methods <u>all</u> exhibit the feature that they successfully compress some types of patterns but expand others. On average, they don't (unless they're badly designed!) make randomly chosen 'typical' patterns either smaller or bigger. However, most pictures, text files, etc, aren't really 'random'. There are patterns which aren't of any value. For example, the text character sequence, 'qgsdxf ftfngt zdplsdesd xotr' isn't very likely to occur in written English. An information storage system which devotes as much storage space to it as to, 'Old Fettercairn tastes great', is being wasteful. Similarly, some characters or symbols occur more often than others or convey less information.

The usefulness of compression techniques comes from matching the technique to the types of pattern you actually want to compress. It essentially removes the redundancy required to encode 'daft' or uninformative patterns. (The daft patterns are then the ones that would come out longer than the original when encoded.) For this reason a variety of compression techniques have been developed, each having its own good points. Here we'll consider a system called *Huffman Coding* after

163

it's inventor. For the sake of illustration we'll use the ancient written language of 'Yargish'. Although it's now rarely used it was once popular amongst the Yargs — a tribe who lived in the hills of Dundee and worked in the tablet mines. (OK, I'm making this up!) The language consisted of just 8 characters — six letters, a 'space', and a punctuation mark. Here I'll represent these characters as, X_1, X_2, ... X_i, ... X_8. By examining lots of Yargish books we find that the relative frequencies or probabilities with which each of these characters occurred were, P_1, P_2, ... P_i, ... P_8. From chapter 5 we can say that the average amount of information (in bits) in a typical Yargish message N characters long will be

$$\langle H \rangle = -N \sum_{i=1}^{8} P_i \log_2 \{P_i\} \qquad \text{...}(18.1)$$

where the angle brackets $\langle \rangle$ are used to indicate that we're talking about an average or typical value. The actual amount of information in a specific message which contains A_1 of the X_1 character, A_2 of the X_2 character, etc, will be

$$H = -\sum_{i=1}^{8} A_i \log_2 \{P_i\} \qquad \text{...}(18.2)$$

An analysis of Yargish reveals that the relative probabilities of the character occurrences are: $P_1 = 0{\cdot}125$, $P_2 = 0{\cdot}5$, $P_3 = 0{\cdot}05$, $P_4 = 0{\cdot}06$, $P_5 = 0{\cdot}055$, $P_6 = 0{\cdot}01$, $P_7 = 0{\cdot}17$, and $P_8 = 0{\cdot}03$. A typical 16-character (i.e. $N = 16$) message would therefore contain $38{\cdot}48$ bits worth of information. However, to be able to indicate 8 distinct symbols we would expect to have to allocate 3 binary bits per symbol to give us the require range of possibilities ($2^3 = 8$). This means that, using a simple fixed-length code like, $X_1 = \%000$, $X_2 = \%001$, $X_3 = \%010$, ... $X_8 = \%111$, we have to send $3 \times 16 = 48$ bits to communicate a 16-character/symbol message. We've already encountered this basic problem. The fixed length coding scheme is *inefficient*. It contains *redundancy* which could be used to help detect and correct errors, but slows down the communication process.

This arises because the symbols/characters used aren't all equally probable. From the above values we can say that the amount of information provided by each individual character's appearance is typically:

$h_1 = -\log_2\{P_1\} = 3$ bits worth for each occurrence of an X_1,

$h_2 = -\log_2\{P_2\} = 1$ bit for each X_2,

and similarly, $h_3 = 4{\cdot}32$, $h_4 = 4{\cdot}06$, $h_5 = 4{\cdot}18$, $h_6 = 6{\cdot}64$, $h_7 = 2{\cdot}26$, and $h_8 = 5{\cdot}05$.

From this result you can see that characters which appear more often only convey a relatively small amount of actual information per occurrence. For example, X_2's which, typically, make up half the symbols in a message only provide 1 bit's worth of information per appearance despite using 3 bits to code. You can also see that rare symbols provide a relatively large amount of information. X_6's typically only occur about once in a hundred symbols, but when they appear they provide 6·64 bits worth of information. This result is interesting because it shows that the fact that a symbol or character might be coded using three binary bits <u>doesn't</u> mean that it always carries just three bits worth of actual information. (However, if all the symbols <u>were</u> equally probable they would each have a P value of 0·125 and an h value of 3. Then the actual number of bits used to represent each symbol <u>would</u> equal the amount of actual information per appearance.)

It is this difference between the actual information content (in bits) and the number of bits required for fixed-length representation — where every symbol is represented by the same number of bits (3 in this example) — which allows us to compress data using the Huffman method. Huffman coding represents each character or symbol by a string of bits whose <u>length</u> varies from character to character. Highly probably characters (like X_2) are represented by short strings. Rare characters (like X_6) are represented by long strings.

The way Huffman codes are produced is shown in fig 18.2. First we list all the codes, X_1 to X_8, along with their relative probabilities. We identify the two symbols which have the lowest probabilities and bring them together, adding their probabilities together. We then treat the gathered pair as a fresh symbol and repeat this process over and over again. In this way we reduce the number of 'branches' by one as we move 'down' from each level to the next. Eventually we will have brought all the symbols or characters together and reached the base of the tree where there's one combination with an accumulated probability of 1. (Assuming, of course, that we haven't missed anything!) The resulting tree will then have as many levels as we have different characters or symbols — 8 in this case.

In principle we may find that three or more candidates at a given level share the lowest accumulated probability values. If this happens we just pick two of them at random and go on. The branch pairs which link a pair of locations on one level with a single location on the one below can be called *Decision Pairs*. We label the two branches of each decision pair with a '1' and a '0'.

165

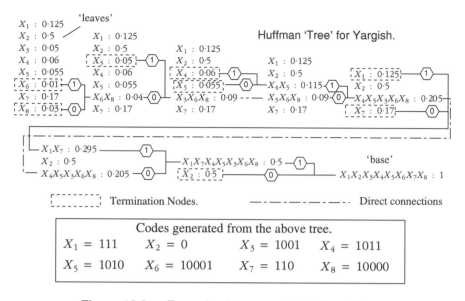

Figure 18.2 Example of the use of a Huffman Tree.

To work out the Huffman code of a given character or symbol we start at the base and follow the branches back up the tree until we find the lowest level where the character appears by itself. The points we arrive at in this way are the *Terminal Locations* (or *Terminal Nodes*) of the characters. Note that there is just one terminal node for each character. The Huffman code can be created by noting in turn the '1'/'0' values which label each of the deciding branches where we have to choose which of two ways to proceed 'up' the tree. The Huffman codes for Yargish are shown in figure 18.2. You should be able to see how these codes are produced by working your way up the tree for yourself and using the above recipe. As you can see, the Huffman code for the most commonly occurring code (X_2) is the shortest, having just one bit. The least common codes (X_6 and X_8) are the longest at five bits each. The codes, X_1 and X_7, which occur with a probability whose value almost equals the value we would get for equiprobable codes ($1/8 = 0.125$) have 3 bits each — i.e. the same number as we need for an equal-length coding system. Consider a specific but fairly typical 16-character Yargish message;

$$X_4X_2X_2X_1X_2X_3X_5X_2X_2X_8X_2X_7X_2X_1X_2X_7$$

Using a plain fixed-length digital code, $X_1 = \%000$, $X_2 = \%001$, $X_3 = \%010$, etc, this produces the result;

$$\%0110010010000010101000010011110011110001000001110$$

which is 16 × 3 = 48 bits long. Using the Huffman code we worked out in figure 18.2 the same message becomes;

%1011001110100110100010000011001110110

which is only 37 bits long.

Using equation 18.2 we can work out the actual amount of information in this particular message. There are $2 \times X_1$'s, so $A_1 = 2$; $8 \times X_2$'s, so $A_2 = 8$; similarly, $A_3 = 1$; $A_4 = 1$; $A_5 = 1$; $A_6 = 0$; $A_7 = 2$; and $A_8 = 1$. Using these values and the probabilities given earlier we get a total information content of $H = 36{\cdot}74$ bits. We worked out earlier that a 'typical' Yargish message 16 characters long would convey 38·48 bits worth of information, so this specific message is slightly on the boring side of typical! Note that the actual number of bits required for Huffman coding is quite close to the actual information content. The equal length coding version, 48 bits long, has about 10 redundant bits in it.

Huffman coded messages are slightly more difficult to decode than conventional equal length systems. When we receive an equal length coded message we can immediately chop it up into parcels of so many bits per symbol. Then we just look up each section in a code table. For Huffman codes we have to proceed as follows.

Begin with the first bit by itself ('1' in the above example). Ask, 'Is this one of our legal Huffman codes?'. If it is, write down the appropriate character and throw the bit away. If not, take the next bit and put it next to one we have. In the example we're using '1' isn't a legal code, so we join the next message bit to it and get '10'. Is this a legal code? No, so take the next message bit and join it on to get '101'. No, so take the next and get '1011'. This is a legal code, for X_4, so we now know that the first character of the message is X_4. Having discovered a legal code we throw away the bits we've accumulated and go on. The next message bit is a '0' — the legal code for X_2. So we know the next character is X_2. Having discovered this, throw the accumulated bit away and find, etc, ... We decode the message by taking the bits in order and accumulating them until we discover a legal pattern. We then recognise that character, throw away the accumulated bits, and continue along the message string until we've accumulated another recognisable legal pattern. In this process we're essentially 'working our way up the tree' until we reach a termination location and we've identified the character sitting at the top of that branch. We then start at the base of the tree and work our way up again.

167

Many modern data compression schemes are *Adaptive*. The above example worked out its Huffman tree codes using the overall probabilities — i.e. using the frequency with which each Yargish character occurs in the whole Yargish language. Given a long <u>specific</u> message we can often do a bit better by using the numbers of times the characters appear <u>in that message</u>. Consider the example of a message which consisted of 16 X_8's. Using the above coding this would become

%100001000010000100001000010000100001000010000100001000010000
100001000010000100001000010000

which is $16 \times 5 = 80$ bits long! In one sense this seems fair, such an unusual/surprising message is likely to carry a lot of information. However, if we'd worked out our Huffman tree using just this message for the probability values we would have said, 'No X_1's, X_2's, ... X_7's, and 16 out of 16 places are X_8's, so $P_1 = P_2 = ... = P_7 = 0$, and $P_8 = 1$ <u>if we just consider this message and ignore the rest of the Yargish language.</u> On this basis we'd have coded X_8 as %0, and all the other characters with longer codes. The coded message would then have been

%0000000000000000

i.e. only 16 bits long. By adapting the coding process according to the <u>message's</u> details we have dramatically improved our ability to compress the number of bits. However, this process has a snag — without the original message how does the receiver of the signal know that a '0' means an X_8?

The advantage of basing the coding on the overall probabilities of various symbols means that all message senders and receivers can agree in advance what coding they're going to use. The adaptive system means each coding system is created specially for that particular message. For it to be decoded we need to provide the receiver or decoder with a 'key'. This is usually done by providing a *Header Table* before the main part of the message. This lists the details of the coding used for this specific message in a pre-agreed form (i.e. <u>not</u> encoded in the special-for-this-message form). Since this part of the message can't be compressed in an adaptive way, and it must give details of all the code patterns, it can be fairly long. As a result, for a short message like the 16–character one we're considering it can end up being longer than the 'main message' itself! Because of this, adaptive encoding is usually pointless for very short messages. Similarly, <u>very</u> long messages tend to use symbols or characters about as frequently as we would expect from their general probability, so adaptive methods may sometimes be a waste of time with long messages.

The main advantage of adaptive compression arises when we have to compress different types of data which have different symbol probabilities. For example, when using a computer we might want to compress both text (ASCII) files and pixel image files. Although both types of file store information as strings of 8-bit bytes the frequencies with which different byte values occur are different for English text and for pretty pictures. A coding scheme fixed to be ideal for text would do a poor job with images, and vice versa. Adaptive encoding means we can use the same computer program to compress/expand all types of files and generally get good results.

Summary

You should now understand how we can use *Data Compression* to remove redundancy from messages and store/transmit them using fewer bits. You should also understand how *Run-Length* and *Huffman* codes are generated and used. You should also know that *Adaptive* coding is useful when we want to compress various types of data, but that it may be better to use a non-adaptive system for specific data types — e.g. text.

Questions

1) Explain the difference between true *Data Compression* and *Data Reduction*.

2) Explain how *Run–Length Encoding* can be used to compress pixel data 'picture' files. What characteristic of pixel data makes this a suitable system for this type of information? When will files encoded in this way come out <u>longer</u> than the initial picture files?

3) A code system consists of four symbols, *A, B, C,* and *D*. These typically occur in messages with the relative probabilities, $P_A = 0{\cdot}2$, $P_B = 0{\cdot}05$, $P_C = 0{\cdot}22$, $P_D = 0{\cdot}53$. How many bits per symbol would be needed to send messages using a *Fixed Length* code system? Use a 'tree' diagram and derive a *Huffman Code* for the symbol set. Calculate how many bits a typical message, containing 512 symbols, would require if encoded in this way. A <u>specific</u> message contains 256 '*A*'s and 256 '*D*'s. How many bits are required to send this message in Huffman coded form? [**2 bits/symbol. Average of 880·64 bits for typical message. 1024 bits for the specific message.**]

Chapter 19

Data thinning

In the last chapter we saw how it's possible to *compress* data in order to save on data storage space or send information more quickly and efficiently. True *data compression* reduces the number of bits or symbols in a message without losing any information. Sometimes data compression still leaves us with an amount of data which is inconveniently large. In such cases it may be necessary to throw away some of the 'less important' information in order to produce a message small enough to communicate or store. Public relations and sales executives don't like to admit this kind of thing, so systems which do this still tend to be called 'data compression' in advertising literature. However, they are more honestly called *Data Thinning* or *Data Reduction* systems. To see how they work we will use two examples. The first is the *JPEG* format for photographic images, the second is the *ATRAC* format used to compress audio data on MiniDiscs. Both of these compression systems – and many others – make use of a specific type of transform called the *Discrete Cosine Transform*. It is therefore a good idea to start by explaining what this is and how it works.

19.1 The Discrete Cosine Transform

Fourier Transformation has already been described at the beginning of Chapter 7. Data thinning is based on an assessment which decides that some details of the data are nominally more or less 'important' than others. For audio or visual data this judgement will depend upon the details of human perception. The rules for this are complex and do vary to some extent from one person to another. However experiments have shown that one of the most effective ways to proceed is to convert signals into some form of frequency spectrum and then discard those frequency components that are 'too small to be missed'.

The *Discrete Cosine Transform* (DCT) is a particular form of Fourier Transform that happens to be convenient in situations where we wish to deal, quickly, with data that is in the form of a long stream of integer values. To understand how it works, and why it is convenient in practice we can begin by considering the example shown in figure 19.1.

170

The lower part of figure 19 shows a digital stream of data of the kind we might expect to be produced from digitising an audio signal. Using the arguments presented in Chapter 7 we could, if we wished, take the entire data stream and Fourier Transform it to obtain a representation of the same information as a series of numbers that indicate the frequency spectrum of the audio pattern.

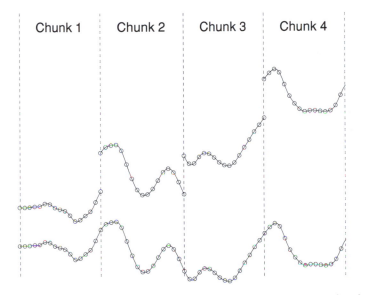

Chunk 1   Chunk 2   Chunk 3   Chunk 4

Figure 19.1 Splitting a long set of samples into discrete 'chunks'

An entire audio recording may contain many millions of sample values, and represent a sound pattern lasting for an hour or more. Fourier Transforming an entire recording of this length would take quite a long time even using a powerful computer. In addition, the transformation process could normally not begin until all the data from the recording had been loaded into the computer's memory. This would also be likely to take a significant time.

Long delays of this kind might be annoying enough in a recording studio, but would be completely unacceptable in a domestic audio player. It is unlikely that a device like a CD or MiniDisc player would ever have become popular if a disc had to be pre-loaded a half an hour or so before the music could be heard! However, breaking the data up into chunks gives us a double advantage. Firstly, each chunk now contains a relatively small number of data points, and hence can be transformed and

171

processed relatively quickly. Secondly, each chunk can be processed without having to wait until later samples have been read. As a result, the processing only produces a short delay as the data is streamed through the processing systems. An additional practical advantage is that the amount of temporary storage (memory) required to hold intermediate values during the calculation will be reduced, thus reducing the complexity and cost of the system. For these reasons it is therefore convenient in practice to break up the data stream into brief, manageable chunks and process these one after another.

A conventional Fourier Transform will take a series of input values and compute two sets of results. Depending on how we wish to represent the process these are either in the form of pairs of amplitude values, A_n and B_n, or magnitude and phase pairs. If we use the same approach as described at the start of Chapter 7 we can say that the A_n values represent the Cosine contributions to the signal pattern and the B_n values represent the Sine contributions. Cosine waves are 'even' or 'symmetric' patterns – i.e. all Cosine waves have the property that

$$\text{Cos}\{x\} = \text{Cos}\{-x\} \qquad \qquad ...(19.1)$$

whatever the value of x. In a similar way, Sine waves are always 'odd' or 'antisymmetric' and are such that

$$\text{Sin}\{x\} = -\text{Sin}\{-x\} \qquad \qquad ...(19.2)$$

The DCT transform differs from a conventional Fourier Transform in two ways. The first difference is that the input and output values of a DCT are generally in the form of integers, nor real or floating point values. The second is that the DCT only computes and uses the Cosine components.

The symmetric properties of Cosine waves means that the DCT can only record or represent the symmetric part of an input pattern. Any antisymmetric patterns or components of the input will be ignored since we have neglected the Sine components which are required to represent them. However, if we look at the typical patterns of the chunks shown in figure 19.1 we can see that the chunks <u>aren't</u> all nicely symmetric shapes. To avoid any unwanted information loss we must therefore do something to the data before performing the DCT in order to take this into account.

The simplest approach is illustrated in figure 19.2. We assume that the origin of the time axis for each chunk is at the start of the chunk. A copy is taken of the chunk pattern (i.e. of the sample values). The copy is then 'reflected' in the time axis and joined to the start of the pattern. The result is a new pattern which is symmetric about time zero for that chunk. Hence each of these modified patterns now only requires a set of cosine components to describe its Fourier Transform.

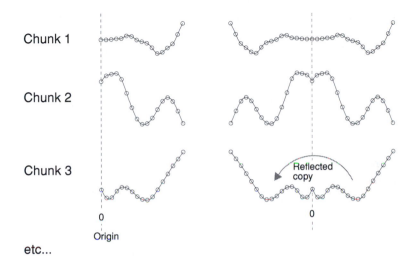

Figure 19.2 Converting chunks into symmetric form

When we now take the DCTs of the modified chunks we find that haven't lost any information by ignoring the sine components of the transform. This is for two reasons. Firstly, the added reflections mean that we have forced any sine components we tried to calculate to always be zero, hence the values are known without having to work them out. Secondly, the new patterns have more sample points than the initial chunks so we have to compute more cosine values to take them all into account. The result is that the DCT generates just as many values as a conventional Fourier Transform, but in a different form. We haven't lost any data by using the DCT, just produced it in a form that may be more convenient.

An alternative way to view the difference between the DCT and a basic Fourier Transform is in terms of the normal assumption, explained in Chapter 7, that the signal pattern is 'periodic'. For the basic transform, this means we assume that the signal's spectrum can be fully described in terms of just the frequencies that will fit an integer number of cycles into the total length (interval) of the set of samples. By adding the reflected copy we now have an extended set of samples, twice as long as the original.

Making the same 'periodic' assumption for this extended set means we now choose a set of frequencies that fit an integer number of cycles into the longer interval covered by the extended set of samples.. This means the frequencies used to specify the spectrum now have an integer number of *half*-cycles in the original signal interval – i.e. we get 'twice as many frequencies' in the required bandwidth. This doubling of the number of frequency components replaces the 'ignored' Sine components to make up the complete amount of data required to represent all the details of the signal.

19.2 JPEG compression

The above explanation assumes a 'linear' stream of data similar to that which we might obtain for a musical signal. JPEG files store data about 2 dimensional still images, usually colour pictures. The name 'JPEG' is an acronym for *Joint Photographic Experts Group*'. Strictly speaking, an actual image file should be referred to as a 'JFIF' (*JPEG File Interchange Format*), but it has become conventional to name the files after the group of people who devised the system. It is also worth noting that there is no single format for JPEG images. Instead the JPEG process is more like a menu of options that may be selected and used as preferred. Here we will just outline a typical JPEG thinning process.

Chapter 18 has already explained how an image can be represented in terms of a bitmap that records a set of values that specify the brightness and colour of a pattern of pixels. A monochrome (black and white) image only requires one number for each pixel since it only records a brightness pattern. A colour picture has to record extra data to indicate the *hue* of each pixel.

The simplest way to record the colour and brightness information would be to have three values, *R*, *G*, and *B*, that indicate the levels of red, green, and blue for each pixel. This approach works but turns out not to be very efficient for a number of reasons. For example, the human eye is much more sensitive to changes in the intensity of green light than either red or blue. In addition the eye has a lower resolution for colour changes than for changes in brightness.

To exploit these human characteristics, the JPEG (along with many other image processing or communication systems) defines the brightness and colour of each pixel in terms of three specific values. One is the *luminance*

174

defined as

$$Y \equiv 0{\cdot}3R + 0{\cdot}59G + 0{\cdot}11B \qquad \qquad \ldots (19.3)$$

This is the equivalent of the 'brightness' of each pixel. The levels of red (R), green (G), and blue (B) are weighted with differing factors. This is to make the resulting value more sensitive to changes in the green level than to the other colours. Hence the luminance value takes the eye's behaviour into account and gives the optimum performance for a given range of data values.

The other two values are *colour difference signals* which we may define as

$$C_B \equiv B - Y \quad ; \quad C_R \equiv R - Y \qquad \ldots (19.4 \& 5)$$

Since we still have three values (Y, C_B, C_R) we can expect to still be able to convey any trio of (R, G, B) values so no information is lost. In principle, all we have done is converted one set of thee numbers per pixel into a different set that represents the same information. However, this new set has some practical advantages which become apparent during the JPEG creation process.

Input set of colour pixel values

R_{11}, G_{11}, B_{11}	R_{12}, G_{12}, B_{12}	R_{13}, G_{13}, B_{13}	R_{14}, G_{14}, B_{14}	...
R_{21}, G_{21}, B_{21}	R_{22}, G_{22}, B_{22}	R_{23}, G_{23}, B_{23}	R_{24}, G_{24}, B_{24}	...
R_{31}, G_{31}, B_{31}	R_{32}, G_{32}, B_{32}	R_{33}, G_{33}, B_{33}	...	
R_{41}, G_{41}, B_{41}	R_{42}, G_{42}, B_{42}	...		
...				

Generated luminance and colour difference signal values

Y_{11}, C_{B11}, C_{R11}	Y_{12}	Y_{13}, C_{B13}, C_{R13}	Y_{14}	...
Y_{21}, C_{B21}, C_{R21}	Y_{22}	Y_{23}, C_{B23}, C_{R23}	Y_{24}	...
Y_{31}, C_{B31}, C_{R31}	Y_{32}	Y_{33}, C_{B33}, C_{R33}	...	
Y_{41}, C_{B41}, C_{R41}	Y_{42},	...		
...				

Figure 19.3 Converting colour pixel values into 4:2:2 format

The JPEG format includes two requirements regarding the size and layout of an image. Firstly, images cannot be be more than 65,536 pixels wide or high. Secondly, the colour difference values must have half the horizontal resolution of the luminance. We can use fig 19.3 to see what this means. The figure represents part of the colour pattern of an initial image in terms of an array of red, green, and blue values. These are then converted into a generated set of luminance and colour difference values. The luminances for each pixel are simply obtained using equation 19.3 and a value is generated for each pixel. The colour difference values obtained using expressions 19.4 and 19.5 are averaged over pairs of input pixels that are horizontally adjacent to one another. This means that some of the original image data is lost and the horizontal colour resolution is reduced by a factor of two.

Since the human eye tends to recognise shapes in terms of brightness changes this loss in colour information usually has surprisingly little effect upon the appearance of the image. It does, however, mean that we have already discarded 40% of the data without significantly degrading the perceived image. This arrangement where we have halved the number of colour values in this way is said to be in '4:2:2' format. This arrangement is required for the input to standard JPEG compression.

The luminance and colour difference values are now divided into three separate sets, for Y_{mn}, C_{Bmn}, C_{Rmn}, and processed separately. Each set of values is DCT processed. Since the process is similar for each of the three sets of values we can now just use the luminance set as an example.

Since we wish to perform a DCT we have to ensure that the pattern is symmetric. Unlike the audio data considered earlier, the data is 2 dimensional, so we have to perform a 2 dimensional transform and start with a pattern that is symmetric in 2 dimensions. This means we have to 'reflect' the data in both the horizontal and vertical directions. We also have to split the total image into 2 dimensional chunks of 8 × 8 values. In practice, this means we have taken a 16 × 8 chunk of pixels from the original image and then will have two side-by-side 8 × 8 luminance chunks and a pair of colour difference chunks to process.

For a given chunk we can represent the luminance data as a set of values, Y_{mn}, where $1 \leqslant m \leqslant 8$, $1 \leqslant n \leqslant 8$. The DCT values can then be computed using the expression

$$Z_{ij} = \frac{C_i C_j}{4} \sum_{m=1}^{8} \sum_{n=1}^{8} Y_{mn} \cos\left\{\frac{\pi i m}{8}\right\} \cos\left\{\frac{\pi j n}{8}\right\}$$

$$\dots (19.6)$$

The i, j values represent the 'spatial frequencies' of variations in the level from pixel to pixel. Here each of these integers can take values from zero to seven to cover all the expected data. As a result we can obtain as many Z_{ij} values for our output 2 dimensional spectrum as there are data points in the original image block. Hence we can expect to obtain a set of spectral values which contain all the information. In practical terms, the i, j values indicate the number of half-cycles of fluctuation in luminance across the block in the horizontal and vertical directions for each component in the spectrum. So, for example, $i = j = 0$ represents the amount of uniform brightness across the block – i.e. the 'd.c.' level of the data.

The coefficients, $C_i C_j$, are chosen so as to normalise the results correctly. This means that $C_j = 1$ if $j = 0$ and $C_j = 1/\sqrt{2}$ otherwise, and similarly for C_i. It is worth noting that the actual calculation specified by equation 19.6 does not actually need to use the 'extra' values which the reflection process creates. We already 'know' these values and, by symmetry, they simply increase the amplitudes of all the frequency components we are calculating by a factor of four. Hence we don't actually need to spend time including them in the computation!

Having divided the image data into blocks and performed the DCT to convert the information into spectral form the next step is to actually 'thin out' the data. To see how this is done, consider fig 19.4. This shows a typical array of Z_{ij} values which we wish to process.

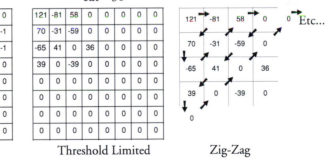

121	-81	58	22	-22	10	-3	0
70	-31	-59	4	-2	0	-5	-1
-65	41	-20	36	22	10	0	-1
39	-6	-39	17	7	5	1	0
-6	-12	-20	-5	-8	0	0	0
-15	11	0	2	1	0	0	0
4	0	-1	0	-1	0	0	0
-1	0	1	0	0	0	0	0

Spectrum

cut = 30

121	-81	58	0	0	0	0	0
70	-31	-59	0	0	0	0	0
-65	41	0	36	0	0	0	0
39	0	-39	0	0	0	0	0
0	0	0	0	0	0	0	0
0	0	0	0	0	0	0	0
0	0	0	0	0	0	0	0
0	0	0	0	0	0	0	0

Threshold Limited

Zig-Zag

Figure 19.4 Thinning the image spectrum data.

Here we start with the spectrum component values which have been obtained from performing a DCT on a block taken from an image. The values shown in the figure are chosen purely for the sake of example. We next choose a *Threshold* level and set any values whose magnitude is less than the threshold to zero. Provided that we choose a high enough threshold this will produce a result where many, if not most, of the values are zero. We can then turn the 2-dimensional pattern into a string by reading through the values along a *Zig-Zag* path. For the data given in the example, this produces a set of 64 values

121, -81, 70, -65, -31, 58, 0, -59, 41, 39, 0, 0, 0, 0, 0, ... etc ...

We can now compress this by using a method such as *Run-Length Encoding*. The result is a set of values

1, 121, 1, -81, 1, 70, 1, -65, 1, -31, 1, 58, 1, 0, 1, -59, 1, 41, 1, 39, 7, 0, 1, 36, 1, -39, 45, 0.

Note that this new set only requires 28 numbers to describe the contents of the entire 2-dimensional pattern. By setting a suitable Threshold and taking a Zig-Zag path we have created a set of data that typically contains large runs of identical values (i.e. zeros), hence we can compress the data from 64 to 28 symbols (integer values in this case) by using Run-Length Encoding.

In most photographic images of natural scenes there is a tendency in most areas of the image for the low frequency components to have significantly greater amplitudes than the higher frequencies. The above process therefore tends to remove the high frequency details of the image. However it only does this in blocks of the image where the high-frequency data is small enough to fall below the chosen threshold. If a block contains large high-frequency components – e.g. at the edge of some artificial object in the picture like a building – these will be above the threshold and will be preserved in the compressed data.

The Zig-Zag path means we tend to cluster together the low-frequency (top left of the spectral pattern as shown) and high-frequency (towards the bottom or right as shown) to mean that we can get the longest possible runs of zeros when we follow the path.

The method therefore tends to adapt itself in a way that allows high-frequency (and hence small-scale 'sharp' details) to be preserved in parts of the image where there are important sharp details, but remove them where the small details have low contrast. As a result the main features of the image are preserved in detail whilst less obvious (to the human eye) details are discarded. Although not explained in detail here, the JPEG process also includes an extra step. This *weights* the spectral components

178

and emphasises some frequencies before the threshold is applied. The purpose of this is to take into account the tendency of the human eye to be more sensitive to some frequencies than others. The effect is to optimise the appearance of the result when it is reconstructed from the JPEG data for a given amount of data thinning.

We may now summarise the JPEG/JFIF creation process in terms of a series of steps

1 Arrange the initial image data into 2-dimensional blocks of integer values.

2 Convert the R, G, B values into Y, C_B, C_R values

3 Perform a 2-dimensional DCT on each block of values

4 'Weight' the values and then apply a Threshold to set the lowest values in each block to zero

5 Convert the 2-dimensional pattern into a Run-Length Encoded series by following a Zig-Zag path.

6 Save the resulting string of values as a JPEG/JFIF file.

The resulting set of data has been Thinned by applying the chosen Threshold level. The higher the level, the more image data we will have lost. This is often under the control of the user. A typical program for performing JPEG/JFIF compression will often provide some form of 'quality setting' to allow the user to make a choice. The higher the chosen Threshold, the smaller the resulting JPEG file will be, and the more detail will have been discarded. The final process is the Run-Length Encoding that packs the remaining information reasonably efficiently by exploiting the runs of zeros that the Threshold tends to create.

19.3 ATRAC audio compression

When compressing or thinning audio data rather than photographic images we may have different criteria for deciding what information might be 'unimportant' and hence may be discarded. However, once this is done, we can apply similar DCT methods to those described in the last two section to process the data. We can therefore concentrate here on the way in which information is selected for thinning or being preserved. As with the JPEG/JFIF there are a number of differing detailed schemes for thinning or compressing audio data.

Here we will tend to concentrate on the ATRAC system (**Adaptive**

179

TRansform **A**coustic **C**oder)which was developed by Sony for their 'MiniDisc' digital recording/replay system. It is worth noting that, in general terms, the systems developed by Philips and others for other applications employ similar methods but differ in detail. As with the JPEG what follows is just an outline explanation designed to explain how the data thinning process works.

As indicated by the previous sections of this chapter, the incoming data is first divided up into short chunks for processing by DCT. However, the process is a modified approach, hence often referred to as MDCT (Modified Discrete Cosine Transform). The incoming stream of digital values is pre-processed by passing the data through a digital filter. This separates the information into three distinct data streams, each containing only the information about a specific frequency band. (Remember that the signal is likely to be stereo audio, so this means two input channels have been pre-processed to obtain *six* streams of digital data.)

The bands chosen for ATRAC are, 0 - 5·5 kHz, 5·5 - 11 kHz, and 11 - 22 kHz. Each band is then MDCT processed to obtain the spectrum it contains. The division into these three bands is a specific feature of ATRAC. It is designed to exploit the fact that the human ear responds differently to high, medium, and low frequency signals and means each band can be processed separately to try and achieve optimum results.

The MDCT process uses 'overlapping' chunks – i.e. each chunk of data samples includes some sample values from the adjacent chunks. This means that each of the resulting chunk spectra contains duplicate copies of some data from the earlier and later chunks. The extra data is then removed to avoid wasting storage space. Although not strictly necessary from the viewpoint of information theory, this process of overlapping and discarding means the results provide a smoother result when thinned data is reassembled.

The reason why MDCT overlapping is useful for musical and speech data can be understood by considering what happens when a sudden transient sound occurs at a time that happens to make it fall across a chunk boundary. The front edge of the transient is processed and thinned by processing one chunk spectrum. The tail of the transient is processed and thinned by processing the next. Since the spectra of the two parts of a quickly changing event are likely to be different there may well be an abrupt change in the pattern which is reconstructed from the thinned data. By making chunks overlap we can ensure that any sudden change in

180

the waveform will always appear in the main part of one chunk. Thus we can avoid discontinuous changes in the way such events are stored and reconstructed. This property is quite important as human hearing is very sensitive to the nuances of brief transient events and changes. Experiments show that these features are a critical part of our ability to recognise sounds and determine their direction of arrival. Overlap is therefore very useful in improving the perceived quality of sound patterns reconstructed from the thinned data, although in strict information theory terms it is not required.

A second feature of ATRAC is that the chosen block lengths for the data chunks is varied depending upon how complex (and loud!) the signal is at any time. When the signal is a simple, periodic pattern large blocks are collected and transformed. When the signal is complex and rapidly changing, the blocks transformed and processed are much shorter. By dividing the input into the three bands it becomes possible to choose their block lengths independently. This tends to help the system optimise the amount of compression (thinning). In practice, the data blocks transformed can be as short as 1·45 ms, or as long as 11·6 ms, depending upon the details of the signal.

Having obtained a set of spectra (one for each band of each audio channel) the next step is to thin the actual data. For ATRAC recordings the data rate must be reduced to 292 kbits/second. This is significantly less than the 1·4 Mbits/second used by conventional audio CDs. As with images the approach is designed to exploit the behaviour of human senses, so we need to consider the properties of human hearing to understand what takes place.

ATRAC encoding (and other forms of audio compression/thinning) makes use of the fact that most musical/spoken signals consist largely of periodic or coherent signals which can be relatively easily described as a combination of a modest number of sine/cosine wave frequency components. It also exploits two characteristics of human hearing called *Masking* and the *Threshold* level. *Masking* is the effect where a loud sound tends to overpower or 'swamp' our ability to hear a quieter tone at a similar frequency. The *Threshold* represents the lowest sound level which can be heard.

In principle, any sound components which are masked or below threshold can't be heard. Hence we can seek to exploit these effects to discard some frequency components without the perceived sound being altered when the resulting thinned spectrum is used to reconstruct the audio waveforms when the recorded music is replayed. In reality, of course, hearing varies from person to person so it it is open to question how effective this may be in a given case. That said, the results when using a modern ATRAC (or similar) system are generally quite convincing and impressive given the large degree of data thinning that has occurred.

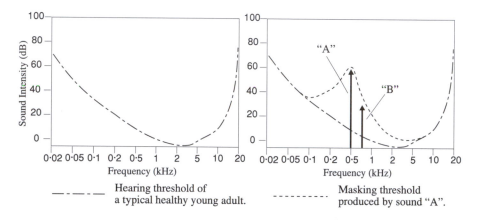

Figure 19.5   Limits of hearing and sound masking effect.

The left-hand graph of figure 19.5 shows the lowest sound intensity which can be heard by a typical healthy young adult. (A male, aged 15–25. Females tend to have have slightly better hearing, older people often have worse hearing.) The line indicates the quietest sounds at various frequencies which are just on the limit of being audible. The sound intensities are quoted in dBs referenced to a level of 10^{-12} W/m^2 which is about the limit of hearing at 2 – 3 kHz. The right-hand graph illustrates *Masking.* In this example a tone of 60 dB (10^{-6} W/m^2) at 500 Hz makes another tone of 30 dB (10^{-9} W/m^2) at 700 Hz 'inaudible'. (Note that in reality, normal human hearing is rather better than the above values would apply. The values given here are purely for the sake of explanation.)

The spectrum of each channel band block is divided into a set of sub-bands as illustrated in figure 19.6. The encoding circuits examine the signal's frequency components in each sub-band. In sub–band 1 the only component, 'A', is below the threshold of audibility, so the ATRAC encoder ignores it. Similarly, there's nothing detectable in sub-band 2. In sub-band 3 there are two signal components, 'B' and 'C'. If 'B' were absent 'C' would be above threshold so the encoder would send

182

information about it on to the recording. However, by comparing the two components the encoder decides that 'C' is masked by the presence of the signal 'B'. So the encoder decides to ignore 'C' but sends information about 'B' on to the MiniDisc.

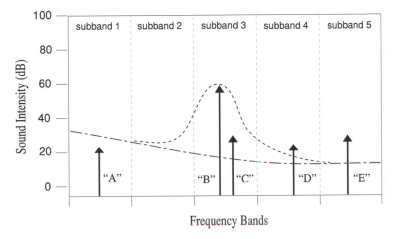

Figure 19.6   Some sub-bands of the frequency spectrum.

In sub-band 4 the signal component, 'D', is high enough for the encoder to decide it is above the threshold and the masking level set by 'B', so information about it should be sent to tape. In sub-band 4 the component 'E' is above threshold, so information about it should be preserved for recording. The ATRAC system scans through all the sub-bands in this way and identifies those signal components which can safely be ignored because they are below the threshold/masking level. The components to be ignored are assigned zero values. Hence as with the JPEG example, if we choose a high enough threshold we can thin out a significant amount of the data and obtain a pattern which will compress efficiently.

The encoder has been programmed to assemble a specific number of output bits per data block which it can use to describe the spectrum of the signal during that frame period. It carries out a process of allocating some output bits to each of the 'audible' components which are then used to describe their amplitudes, phases, and frequencies. This is a *requantising* process where we can, again, reduce the number of bits required by lowering the precision of components the system judges to be of lesser importance. The most powerful components tend to be allocated more bits so that they can be specified more precisely.

183

As with CD, ATRAC/MiniDisc recorders also employ data interleaving, Eight-to-Fourteen Modulation, parity bit error detection and correction, CIRC encoding, etc, to try to avoid problems due to data bits being lost during replay. Although some details are modified, the MiniDisc is similar in many ways to CD. Here we can ignore these similarities as we are only interested in using it as an example of data thinning.

The success of the thinning system can be seen at the most simple level in the fact that, despite the similarities, the actual MiniDisc is physically much smaller than a normal CD. A MiniDisc typically thins the number of bits which have to be stored by approximately a factor of five whilst preserving a sound quality broadly similar to the uncompressed CD. Indeed, some modern MiniDisc recorders use ADC/DACs with more than 16 bits/sample (typically 20 bits/sample) so in some ways they can be argued to be better than a conventional CD.

Summary

You should now know how the JPEG and ATRAC systems for *Data Reduction* or *Data Thinning* work. That each exploits features of the situation where it is applied — the properties of human vision and hearing — to identify and discard information that is judged to be relatively 'unimportant'. That each optimises its output in an *adaptive* way to the signal details. You should now also understand that data reduction techniques can be useful provided that the process of deciding which parts of the signal to discard is performed in a manner appropriate to the context — i.e. it depends upon the nature of the signal and the use to which it is to be put. It should also be clear that the success of data thinning depends critically on how well the system avoids removing information about signal details which do, in fact, matter!

Chapter 20

Chaos rules!

Engineers generally prefer to use circuits and systems which behave predictably. For example, when designing digital logic they tend to use circuits that perform well-defined functions like **AND, OR,** or **NOT.** A careful search through the catalogues of chip manufacturers won't uncover any **PROBABLY, SOMETIMES,** or **WHYNOT** gates! Similarly, when we buy a new watch we expect it to keep 'good time'. The hands should move or the displayed number change at regular intervals. A clock whose hands moved unpredictably faster or slower, perhaps even sometimes going backwards, wouldn't be much use — except perhaps to someone producing a railway timetable...

Most of the simple signal generators used in engineering and science produce *periodic* output patterns like sinewaves or squarewaves of a well-defined frequency. We also tend to analyse more complex signals in terms of combinations of sets of periodic signals — e.g. Fourier analysis which represents signals as patterns of sinewaves. Despite this, there are signals which vary in a very different way.

The most familiar non-periodic signals are *random noise*, and we spent some time considering noise and its effects at the beginning of this book. In this chapter we'll consider a new sort of signal and signal-source called *Chaotic*. Both random noise and chaotic signals/oscillators have important uses in special applications like secret or *Encrypted* messages. We'll be examining secret messages in the next chapter. First we need to discover some of the basic properties of chaotic signals and the systems which create them.

20.1 Driven nonlinear systems and bifurcations

For a system to be able to produce a chaotic signal it has to exhibit some kind of *Nonlinearity* in its behaviour. A simple example of a nonlinear electronic device is a diode. The current passing through a diode isn't simply proportional to the voltage across it. Diodes do not obey Ohm's Law, unlike resistors they have a nonlinear current–voltage relationship.

185

Another requirement for a system to behave in a chaotic way is that it has to have some kind of 'memory' built into it so that it's behaviour <u>now</u> depends upon what happened to it a while ago. Note that although these general conditions are required, they don't <u>guarantee</u> that a system will show chaotic behaviour.

One of the simplest kinds of electronic system which fits the bill is illustrated in figure 20.1. The resistors, R_1 and R_2, inductors, L_1 and L_2, and capacitor, C, in this circuit make what RF and microwave engineers would called a *Lumped Element Network* version of a very short length of *Transmission Line*. (Here the term, 'lumped element', means 'made from a set of distinct components' rather than an actual length of cable or line.) The network connects a *Varactor Diode* to a pair of signal sources, V_{ac} and V_d. A varactor is a capacitor whose capacitance varies with the applied voltage. For reasons we won't bother with here some diodes, when reverse biassed, have this property. Hence diodes of this type are called varactor diodes. In general, the varactor's capacitance tends to fall rapidly as the applied voltage is increased.

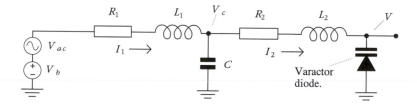

Figure 20.1   Varactor diode driven via a simple RCL circuit.

An inductor will store energy in the magnetic field set up by the current flowing through it. Similarly, a capacitor will store energy in the electrostatic field between its plates when charged by an applied voltage. The above circuit has two inductors and two capacitors (including the varactor), hence it contains four elements which are able to store some signal energy. This ability to store patterns of energy gives the system its 'memory' of what has happened in the recent past. In effect, the system can 'remember' four values — two inductor currents and two capacitor voltages — which are a record of what has been happening recently.

In this system the nonlinearity is provided by the capacitance/voltage behaviour of the varactor. Unlike a normal capacitor, the capacitance of a varactor can be specified in two ways. To see why, let's go back to the basic definition of capacitance. For a fixed-value capacitor we can say that an

186

applied voltage, *V*, will cause the capacitor to store an amount of charge

$$Q = VC \qquad \qquad ...(20.1)$$

where *C* is the value of the capacitance. Alternatively, we can say that changing the applied voltage by a small amount, ΔV, will alter the stored charge by an amount

$$\Delta Q = \Delta V C \qquad \qquad ...(20.2)$$

We can use either of these expressions to define the capacitor's value. Equation 20.1 gives us the *Static* capacitance value. Equation 20.2 gives us the *Dynamic* or *Small Signal* capacitance value. For a normal capacitor these values are identical, so can use the two equations and values interchangeably. However, the static and small signals values are usually <u>different</u> for a varactor as we can see from the following argument.

Consider now what happens when we change the applied voltage on a varactor from a level, *V*, to $V + \Delta V$. We can say that the change in the stored charge will be

$$\Delta Q = \Delta V C\{V\} \qquad \qquad ...(20.3)$$

where $C\{V\}$ is the varactor's <u>small signal (dynamic) capacitance</u> at the voltage *V*. (We'll assume ΔV is very small so $C\{V + \Delta V\} \approx C\{V\}$.) We can work out the total charge stored in the varactor when the applied voltage is *V* by starting at zero volts and integrating expression 20.3 up to *V* volts. This gives us

$$Q\{V\} = \int_0^V C\{V\}\, dV \qquad \qquad ...(20.4)$$

From the static definition of capacitance we can say that the varactor's static value at *V* will be

$$C'\{V\} \equiv \frac{Q\{V\}}{V} = \frac{1}{V}\int_0^V C\{V\}\, dV \qquad \qquad ...(20.5)$$

This result gives a <u>static capacitance</u> value of $C'\{V\}$ which generally differs from $C\{V\}$ when the capacitance varies with the applied voltage. When considering varactors it is therefore important to keep this difference between the small signal and static (d.c.) values in mind. Most data on varactor diodes show how the small signal capacitance varies with the applied voltage since this is what most rf/microwave engineers are interested in. We'll therefore use the small signal or dynamic value unless otherwise specified.

The behaviour of the varactor + circuit system depends upon the choice of component values and the details of the applied signals. Real varactors have very small capacitances — typically less than 100 pF — so for our example we'll use 'artificial varactor' shown in figure 20.2a. This circuit mimics a varactor, but has a much larger small-signal capacitance whose voltage dependence is shown in figure 20.2b. (Anyone who wants to know more about this is welcome to read *Electronics World,* June 1991, pages 467–72, but you don't have to read it to understand this chapter!) This lets the system work at 'audio' frequencies rather than at RF/microwave frequencies.

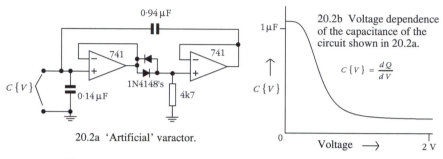

Figure 20.2   An example of a circuit which provides a nonlinear voltage–capacitance relationship.

For our purposes, the precise details of how this artificial varactor arrangement works don't matter. We can just concentrate on what happens when we apply an input signal to the network of

$$V_d + V_{ac} \operatorname{Sin}\{2\pi f t\} \qquad \dots (20.6)$$

which is a combination of a d.c. level, V_d, and a sinewave of amplitude, V_{ac}, and frequency, f.

For our illustration we'll choose $f = 1300$ Hz and $V_{ac} = 3.5$ V, and use component values of $L_1 = 3.24$ mH, $L_2 = 3.38$ mH, $R_1 = 105$ Ω, $R_2 = 4$ Ω, and $C = 2.03$ μF. There is nothing 'magic' about these odd values. They're simply the values of the components picked out of the boxes when this circuit was soldered together! Slightly different values would give slightly different results, but the same overall pattern of behaviour.

We can now examine what this nonlinear system does as we slowly increase the applied d.c. voltage, starting at $V_d = 0$. Figure 20.3 illustrates the results of doing this. The top graph of figure 20.3 shows the input sinewave. The graph immediately below it shows how the resulting voltage across the varactor varies with time when the d.c. level is zero (i.e. $V_b = 0$). Comparing these top two patterns we can see that their shapes are very different, but that both waveforms repeat with a *period*, $T = 1/f$. The next graph down shows the varactor voltage waveform when we apply a small d.c. level, $V_b = 0.08$ V, which is added to the input sinewave. Now the period of this 'output' wave, T_0, is twice that of the input. Increasing the d.c. level slightly, to $V_b = 0.11$ V, increases the period of the output to $T_0 = 4T$.

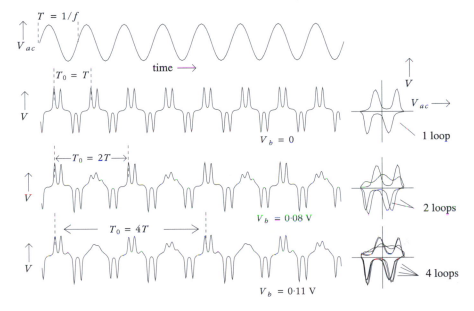

Figure 20.3   Effects on output waveform of varying the d.c. input level.

This process is called *Period Doubling* for fairly obvious reasons. If we explore the behaviour of the circuit carefully we find that it occurs at a series of well-defined *Threshold* voltages, V_1, V_2, V_3, etc. When $0 \leqslant V_b < V_1$, $T_0 = T$; when $V_1 \leqslant V_b < V_2$, $T_0 = 2T$; when $V_2 \leqslant V_b < V_3$, $T_0 = 4T$; when $V_3 \leqslant V_b < V_4$, $T_0 = 8T$; etc... As a result, when we apply a d.c. level great enough for n period doublings to have occurred, we find that the output waveform shape only repeats itself after a period, $T_0 = 2^n T$.

As a result of these doublings the output signal can have a repeat period which is much longer than the period of the *Driving* or *Pump* signal (the

189

input sinewave). For example, when we raise V_b to get 20 period doublings, an input at a frequency of 1300 Hz (T = 0·769 milliseconds) will produce an output waveform which only repeats itself every $2^{20} \times 0·769$ milliseconds = 13·4 minutes!

Consider now the voltage intervals between successive doubling thresholds. If it is always true that $|V_{n+2} - V_{n+1}| < |V_{n+1} - V_n|$ then the doublings become more and more closely spaced as we increase the voltage. When this is the case it becomes possible to pass through an infinite number of doublings while V_b remains finite. This is called a *Cascade to Chaos*. The output signal now only repeats itself after a time interval of $2^\infty T$ = ∞, i.e. the output waveform shape never repeats itself. It is therefore a non-periodic waveform. Such an output is said to be *Chaotic*. Just like random noise we can't predict what it will do later unless we know all the details of the system which is producing it.

Systems which are behaving chaotically exhibit a property called *Sensitivity to Initial Conditions*. Although their behaviour is *Deterministic* — i.e. we know the rules or equations which determine the behaviour from moment to moment — we can't say what they will do in the far future unless we know with absolute accuracy all of the component values, voltages, and currents at some time. Any errors in our values, however small, will eventually mean our predictions are totally wrong. For the same reason it's impossible to make two chaotic systems which behave identically since we can never find pairs of resistors, etc, which are absolutely identical. The processes which generate weather are chaotic, hence the impossibility of making good long range forecasts!

20.2 Chaotic oscillators

The system we've looked at so far is driven with a combination of a d.c. level (V_b) and an input sinewave. Figure 20.4 shows how it is possible to make the system's chaotic oscillations self-sustaining without the need for an input sinewave. This *Chaotic Oscillator* consists of the nonlinear system we've already considered plus an extra LC section and a *Schmitt Trigger*. The output from the trigger circuit is then fed back to the input of the system and used to drive it's behaviour. The Schmitt trigger acts as a high-gain amplifier which produces a 'squared off' version of the voltage on C_3. The Schmitt circuit also distorts the signal (more nonlinearity!) and exhibits *Hysteresis*. For our purposes the details of how a Schmitt trigger works don't really matter. We'll just look at what happens when we build and use the above circuit.

190

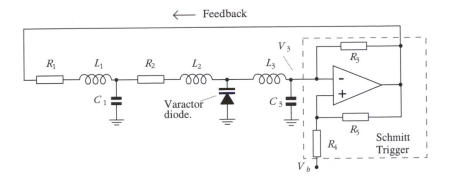

Figure 20.4   Chaotic 'phase shift oscillator'.

Circuits of the same general form as 20.4 are often used as 'clocks' or oscillators to produce regular — i.e. periodic — waveforms. If we replace the varactor with an ordinary fixed-value capacitor the system becomes a conventional *Phase Shift Oscillator*. As an illustration of this, figure 20.5 shows how the voltage on C_3 would vary with time if we make all three capacitors have the same fixed values. (i.e. we replace the varactor with an ordinary capacitor.)

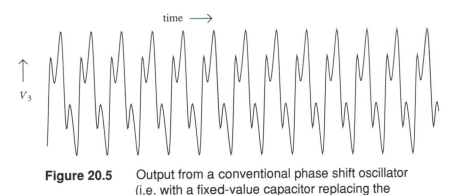

Figure 20.5   Output from a conventional phase shift oscillator (i.e. with a fixed-value capacitor replacing the varactor shown in figure 20.4).

The voltages and currents then oscillate in a simple periodic way, with a periodic time set by the values of the inductors and capacitors we've used. However, using our <u>varactor</u> as the middle capacitor, the circuit shown in figure 20.4 produces output of the general form illustrated in figure 20.6.

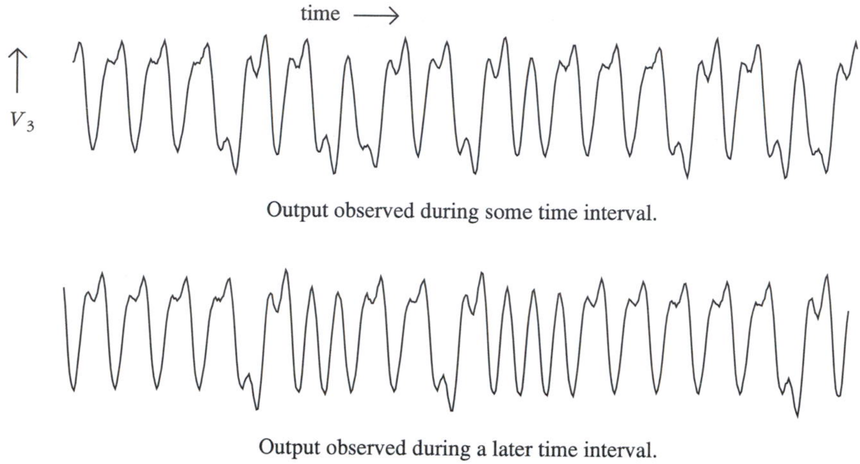

time $\longrightarrow$

Output observed during some time interval.

Output observed during a later time interval.

Figure 20.6   Output from chaotic phase shift oscillator (i.e. with varactor).

Now the oscillations can be seen to 'jitter' or vary unpredictably from cycle to cycle. Although from time to time the oscillation appears to settle down into a repeating pattern, it eventually changes into a pattern we've not seen before. The voltages and currents in the circuit vary chaotically from moment to moment. The behaviour of the system depends upon the exact values of the components used. The waveforms shown in figure 20.6 were produced by a system whose varactor components are as shown in figure 20.2, and $R_1 = 130\,\Omega$, $R_2 = 4\,\Omega$, $R_3 = 10\,\text{k}\Omega$, $R_4 = 10\,\text{k}\Omega$, $R_5 = 10\,\text{k}\Omega$, $L_1 = 3{\cdot}24\,\text{mH}$, $L_2 = 3{\cdot}38\,\text{mH}$, $L_3 = 3{\cdot}5\,\text{mH}$, $C_1 = 2{\cdot}03\,\mu\text{F}$, $C_3 = 2\,\mu\text{F}$, $V_b = 0{\cdot}1\,\text{V}$, with a Schmitt Trigger whose output is $\pm3{\cdot}5\,\text{V}$.

Many different types of circuit have been developed which behave as chaotic oscillators. They all have to provide the same set of basic features: the system must contain one or more nonlinear elements; there must be some gain to boost the signal and counteract any losses; and feedback is applied so that the boosted output is used to drive the system into further oscillations. It is common for systems to employ hysteresis because this produces a 'folding' action where one input level can give either of two output levels depending on the system's recent history. (This is another 'memory' mechanism as well as a source of extra nonlinearity.)

20.3 Noise generators

It is surprisingly easy to make a digital 'random number' generator. Figure 20.7a shows an example of a *Maximal Length* shift register circuit which can be used to produce an apparently randomly varying sequence of output '1's and '0's.

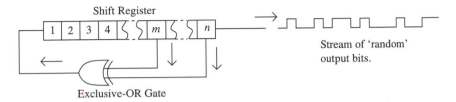

Figure 20.7a   Maximal length digital pseudo-random noise generator.

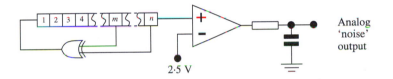

Figure 20.7b   Analog noise generator based on a digital process.

In the analog systems we've looked at up until now signal information/energy was held by capacitors and inductors. The pattern of current/voltage values remembered at any moment is said to be the system's *State* at that instant. In the digital examples shown in 20.7 information about the state of the system is stored as a pattern of bits in a shift register n bits long. The feedback and nonlinearity are both provided by an *Exclusive-OR* gate which takes its inputs from two of the register locations and drives the 'lowest' or first location. If we now repeatedly step the bits along the register we generate an apparently random sequence of output digits. Systems like this are often used as simple noise generators. As illustrated in figure 20.7b they can also be used as part of a circuit which produces an analog voltage which varies in an apparently random manner.

Although often regarded as noise generators, these digital systems cannot actually produce true random noise. This is because — like all digital systems with finite memory capacities — they are *Finite State Machines*.

193

Given n-bit storage patterns we can only store 2^n patterns of information (or *states*). As a result, if we drive the shift register with a shift clock whose period is T we find that the output pattern <u>must</u> repeat after a time of, at most, $T_0 = 2^n T$. This is because the system will have then 'cycled through' all the possible bit patterns it can store and must then repeat a previous state. The digital system therefore behaves like an analog system which has undergone a <u>finite</u> number of period doublings. We can increase the repeat period, $2^n T$, by using a longer register but we can't ever make the repeat period infinite.

In fact there's always at least one 'inaccessible' state. For a shift-register system of the type shown this is the 'All bits 0' state. If the system starts in this state it gets 'stuck' there and never moves on to any other. Since the system's step-by-step behaviour is reversible this also means it can never <u>reach</u> this state if it is oscillating. There are therefore only $(2^n - 1)$ *accessible* states for the shift register to pass through during its 'random' sequence, so the maximal length of time is strictly $T_0 = (2^n - 1) T$, not $2^n T$. It is important in practice to ensure that the system isn't in the inaccessible state when it is switched on, otherwise the oscillation process 'won't start'. This is another reason why the analog system illustrated in figure 20.4 includes a Schmitt Trigger. The trigger prevents the system from sitting in the 'all currents and voltages zero' state when it is switched on.

Some typical register length and *Tapping* values (the values of m and n) are: $(m; n) = (7; 6)$ giving $T_0 = 127T$, $(15; 14)$ giving $T_0 = 32,767T$, and $(31; 28)$ giving $T_0 = 2,147,483,647T$. This last choice would mean that driving the $31/28$ system with a clock rate of $T = 1$ millisecond produces an output which only repeats itself after 24·8 <u>days</u>! As a result, if we only observe or use the sequence of values this system produces for a day or two, it's output can be regarded as 'indistinguishable from random' for most practical purposes.

Nonlinear analog systems which have undergone a finite number of period doublings can be said to oscillate in a *semi-chaotic* manner and produce a semi-chaotic output signal. Just like the maximal-length digital system their output can appear random if it's only observed over a time interval less than $2^n T$. However, when observed for longer than this, the repetitive non-random noise behaviour becomes clear. These repetitive properties mean that finite state and semi-chaotic systems produce what is called *Pseudo-Random Noise*. It looks a bit like noise, but reveals its periodic behaviour if you wait long enough.

194

Chaos rules!

True chaotic systems can make ideal noise generators since their output never repeats itself. A random sequence can be generated in various ways. For example, we can present the 'squared-off' output from the Schmitt Trigger to a counter/timer circuit. This repeatedly measures the time taken for the chaotic signal to oscillate through a given number of cycles (e.g. 32 cycles). Since the chaotic signal oscillation jitters unpredictably, the sequence of time values produced vary in a random manner, hence giving us an output series of random numbers. In the next chapter we'll see how sequences of random numbers like this can be used for information encryption.

Summary

You should now know that nonlinear systems can be used to produce either *Chaotic* or *Semi-Chaotic* output signals. That chaotic signals share with natural noise the property that they are unpredictable and never repeat themselves. You should also now see the relationship between digital *Finite State Systems* which generate *Pseudo-Random* output and semi-chaotic oscillations — both of which <u>do</u> repeat after a specific time. You should now understand that chaotic oscillation requires the system to include a combination of *Nonlinearity, Feedback,* and some way for the system to store information/energy patterns which depend upon the system's *State* at previous times.

Questions

1) Draw a diagram of an example of an analog *Chaotic Oscillator* and say what features are essential for it to be able to behave chaotically.

2) Explain the term *Period Doubling.* What is the essential difference between *Semi-Chaotic* and *Chaotic* behaviour? Draw a diagram of a digital pseudo-random number generator. Why is it impossible for such a system to generate 'true' random noise?

3) A digital system uses a 22-bit shift register and an *Exclusive-OR* gate to generate a *maximal length* pseudo-random bit sequence. The system is clocked at 100 kHz. How long is the time interval before the sequence repeats itself? **[41·9 seconds.]**

Chapter 21

Spies and secret messages

21.1 Substitution codes

Usually we want to transmit information as quickly and efficiently as possible. This means using systems which are simple and don't waste channel capacity. There are, however, times when the priorities are different. These arise when we're concerned with the *Secure* transmission of information. This requires us to devise methods of encoding and transmission which prevent information from 'falling into the wrong hands'. Alternatively, we may find ourself in the situation of needing to discover information which the sender doesn't want us to have. We then have to set about decoding a message which has been encoded in manner designed to make our task virtually impossible!

Espionage and counter-espionage provides some excellent (and very interesting!) examples of the basic methods of encryption and how codes can be 'broken' by various means. One of the most common systems during the 20th century uses five-digit *Code Groups*.

The system relies upon some form of *Code Book* which lists all the words which are likely to be needed. Each word in the book is linked to a five-digit number, and (of course) the numbers in the book are in a suitably randomised order. Using this system we might encode a message

THE	RAIN	IN	SPAIN	as
24397	34651	50904	18253	

This is a simple *Substitution Code* where each word is replaced by a specific *Code Group* of five digits. We could now send the signal 24397 34651 50904 18253 to represent the message in an encoded form. Using another copy of the book this stream of digits could be *Decoded* back into plain text by the person receiving the signal. To anyone else, the signal just looks like a stream of numbers.

Now, in the English language, the letter 'e' is far more common than 'q', and 'th' is a lot more common than 'zq'. Similarly, the word 'the' is more common than 'zebra', and combinations like 'nice day' are more common than 'tree fish'. Hence, given a reasonably long encoded message, an eavesdropper could attempt to unravel the code simply by seeing which groups of numbers occurred most often and by looking for patterns. This use of relative probabilities is called *Entropic Attack* because the expressions for the amount of information per symbol are linked to symbol probabilities by expressions similar to those used for *Entropy* in thermodynamics. (Have a look at the equations in Chapter 5 and compare them with ones for 'real' entropy in a thermodynamics book.)

In a case like the British listening to German transmissions during the war, or GCHQ listening during the 'cold war' to the USSR's embassy in London, a large number of code group strings can be collected. Using entropy and relative probability methods based on the ideas outlined in earlier chapters it would only be a matter of time before a simple substitution code was 'broken' and messages would become easily readable by the eavesdropper. The only way to prevent entropic attack on encrypted messages from being successful it to keep changing the code book. For example, 'The' might equal '24379' one week, and then be changed to '19935' in the book used for the next week's messages. Regular changes in the chosen code groups make it difficult for a codebreaker to collect enough messages using the same code for entropic attack to work. Unfortunately, even a single long message may be enough to identify many of the most common words. Its surprisingly easy to break substitution codes — especially if you have some idea of the kinds of things which are likely to be in the message.

21.2 One-time pads

In order to ensure that entropic methods can't succeed it's necessary to 'change the code book' after EVERY code group — i.e. after every word. This results in a string of code groups which are essentially random in appearance and makes entropic analysis useless. The only problem with this method is that the information become so well encrypted that it may be a problem ensuring that the <u>intended</u> recipient is able to decode the message!

For spies, the traditional method for achieving this 'randomised' encryption was the *One-Time Pad*. This consists of a pad of paper sheets. Each sheet has printed on it a string of five-digit random numbers. For security only <u>two</u> copies of any pad of numbers are printed. One is given to the 'spy' and the other to whoever is meant to be receiving his messages.

The sender (i.e. the spy) first encodes the message as before into a string of five-digit code groups. These are written, in order, on the top sheet of the pad. The first code group is added to the first random number, the second code group added to the second random number, and so on to the end of the message. The numbers are added without performing a carry forwards. (This ensures all the results are also just five digits long.)

Using the example given above this gives something like

THE	RAIN	IN	SPAIN	the message
24397	34651	50904	18253	from code book
47656	23311	93705	49910	from pad
61943	57962	43609	57163	combination

The codes transmitted are 61943, 57962, etc. If the numbers on the pad are a random sequence then the transmitted code groups will be truly randomised. After being used just once the sheet of the pad must be destroyed and the next sheet used for the next message. Entropic analysis will be unable to decipher the encrypted series of digits unless the codebreakers have some information about the numbers on the pad.

Used correctly, the one-time pad system is unbreakable as the transmitted numbers are genuinely randomised. To decode a message you must know the random sequence used to encode it. The weaknesses of the one time pad system are a direct result of the method it uses to ensure message security. Firstly, every message requires a new sequence of randomly varying numbers. Secondly, a copy of this random sequence has to be delivered to the sender without being intercepted (and copied) by anyone else. For a spy this system also has the additional unpleasant disadvantage that being caught with a pad of such random numbers has, in itself, often been used as proof that the possessor is a spy!

The drawbacks of the one-time pad system become a real nuisance when you want to use it to send a large number of frequent, long, messages. During the second World War and afterwards the USSR used one time pads for virtually all of their 'secure' messages. This required vast numbers of pads to be shipped around, each pad being destroyed after just one use. During the war it was particularly likely that pads would be destroyed or copied in transit. To ease these problems they decided to use each pad more than once. As a result, although the numbers in any individual message would seem random, those of a collection or *Ensemble* of messages could be analysed to reveal a pattern. This lead to western intelligence agencies being able to break the encryption of a number of Russian messages.

21.3 Mechanical 'randomising' algorithms

The problems of one-time pads have been known for many years and an alternative method had been available since the 1920s. This replaced the pad with a mechanical system which modern information theorists would describe as providing an *Algorithm* to generate a string of codes which act as if they were random. In this mechanical system, each letter was replaced by a different one according to the positions of a set of *Encoding Wheels* in a coding machine. After each letter one or other of the wheels is rotated by a set amount to change or 'scramble' the code.

First arrangement — Input 'B' leads to an output 'D'.

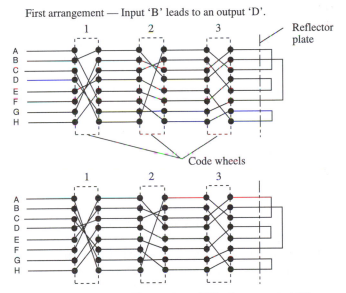

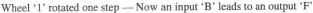

Wheel '1' rotated one step — Now an input 'B' leads to an output 'F'

Figure 21.1   Mechanical encryption system.

This system is the one used by the German forces during the second World War. The British operation to break this code is the, now, much fabled *Ultra* story. The German code machines were called *Enigma*. Messages based upon the use of these code machines <u>can</u> be broken using entropic methods. This is because the codes are not truly random as the rotation (and patterns) of the wheels progress in a set way. Given a reasonable number of long messages (in the jargon of codebreakers, this is called 'having enough traffic') it is possible to break the code. The basic structure of an Enigma code-machine is shown in figure 21.1. Here we'll use this example to show how a *Deterministic* system can produce apparently randomised, encrypted signals. Note, however, that the following description has been simplified to bring out the main points. The actual system was rather better and more complex than it might seem from what follows.

The machine contains three wheels or 'rotors' with a set of metal contacts around each side near the rim. Each contact on one face is wired to another on the opposite face, but the connections are made in a fairly randomised way. The above example shows three rotors with different patterns of connection. Wires are taken from the keys of a typewriter to contacts which touch those of one side of the 'first' rotor. Similarly, wires connect the rotors together.

The signal from a particular key (e.g. the 'B' shown) passes through the rotors and wiring to emerge on one of the wires touching the 'far' face of the third rotor. Here there are a set of wires which make jumbled connections between pairs of rotor contacts. This 'reflects' the signal back through the series of wheels and hence has often been called a reflector plate. The signal then passes back through the rotors by a different path, and emerges on the wire connected to a different letter. (In the case shown above, 'B' emerges as 'D'.)

If the rotors were not allowed to move, the encoded message would be easily broken. However, the machine is arranged to move a rotor after every letter. Here, after encrypting our initial letter ('B' as a 'D') the rotor #1 is rotated by one step. If the next letter in the message is also 'B' it is *NOT* coded as another 'D' but as 'F' because the wiring has changed due to the rotor movement. After each letter the rotor is moved and we have a 'new' code for every letter. When the first rotor has made a complete turn the second rotor is moved one place and then left whilst the first makes another rotation, and so on...

By using a *Reflector Plate* we can double the number of wheel connections each signal must pass through. At first glance this may seem to provide a considerable increase in the amount of jumbling produced, but this isn't really true. Take the illustrated example using three wheels. The signal passes through a total of six wheel connections, but the jumbling produced by the fourth is related to that produced by the second as it is physically the same wheel. Similarly, the second is related to the fifth and first to sixth. This doesn't matter if we are sending messages which are only a few letters long. However, as we keep rotating the wheels for a longer message we will begin re-using connections which have already been used in the other direction.

If each rotor has, say, 32 contacts we appear to have $32 \times 32 \times 32$ possible codes. (Since this represents the total number of physical arrangements we can find for a given set of wheels, this number can't be increased by using a reflector plate.) Hence we could expect to send a message containing 32,768 letters (including spaces, commas, etc, as letters) before having to repeat a code setting. In English this is equivalent to a message around 8,000 words long. This number can be increased if we have some spare code wheels which we can use to replace our initial trio when we approach 32,000 letters.

In effect, we have a machine which can replace the one-time pad and mechanically recode each letter of our message in a different way. This system is very easy to use. It also has the property of being *Self-Inverse* — i.e. if we set up an identical machine and feed in the 'DF' of the encoded message we get the output 'BB'. The same machine can therefore be used both for encrypting and reading the messages! The machine, with a particular set of starting positions for the rotors, gives us a mechanical version of a particular one time pad. We can 'change the pad' by changing the rotors or altering their starting positions. During the 2nd World War, the Germans added a few extra gadgets to make the encryption more complex, but the basic system remained as described. Each rotor had a ring attached to it which was inscribed with the letters of the alphabet. These rings could be slid around to take various orientations with respect to a mark on the rim of the rotor. The machine was then used as follows:

The operator would look up in a standard book which rotors were to be used on that day. The book would also tell him which ring letter to align next to the mark on each rotor rim. The receiver operator would use another copy of the book and set his rotors and rims in the same way. The transmitting operator would then choose for himself the starting positions he wished to use for his rotors when sending the bulk of the actual message. This was necessary to avoid all the messages on a given day having the same encryption algorithm. However, the transmitter now had to indicate to the receiver which starting positions he had chosen. This need for an *Indicator* in the message is necessary unless the codes are to be totally pre-determined by a book or table of instructions (copies of which may fall into the 'wrong hands'). It is a potential source of weakness.

201

In the German system, the operator could identify the starting positions chosen for the message by telling the receiver which ring letters could be seen through three small openings in the machine when the rotors were set up ready to start. This was done by setting the rotors in the day's standard starting arrangement and sending the three letters which indicated the ring settings he'd chosen for the rest of the message. To make sure there was no mistake these letters were then repeated. The rotors were then re-set with the appropriate letters showing and the actual message sent.

This method of indication is a poor one. It relies on assuming that the basic encryption is so good that it is effectively unbreakable. In reality, the Enigma system can be broken given information about how it works. Various accounts of how this was done have been published in recent years. Perhaps the best explanations are in *Hodges; The Enigma of Intelligence* (a scientific biography of Alan Turing, published by Hutchinson, London, 1983) and *Hinsley and Stripp; The Code Breakers* (a collection of memoirs from people who worked at Bletchley Park during the second World War, published by Oxford University Press, 1993). The basic method was as follows:

In Enigma codes it is impossible for any letter to be encoded as itself. This is because of the reflector plate, which always returns a typed signal back through the wheels on a path different to the one it followed on its way towards the plate. Hence the initial six letters 'ADGSFH' in a message must stand for an indicator where the first letter cannot be either 'A' or 'S', the second cannot be 'D' or 'F' and the third cannot be 'G' or 'H'. Importantly, all the traffic on a given day uses the same code scheme for those first six letters. Using these two pieces of information it is possible to try various possible codes until one is found where, for every message, the first six letters are a group of three repeated and none are the same as the corresponding encrypted letters.

The Allies were fortunate in getting some information shortly after the start of the war on the wiring of the standard rotors and the method of operation. (Some of the details of this remain secret, although it is clear that considerable help was provided by the Polish intelligence service.) Hence once the indicators were broken, all that day's traffic could be de-encrypted.

21.4 Electronic encryption

The Enigma system is an example of the use of an algorithm for encoding messages. Modern systems exploit the power of large, fast, digital computers. This enables them to carry out far more complex encryption schemes, and produce codes which are harder to crack. It remains true, however, that <u>any</u> *deterministic* method of coding can — in principle — be broken. The main object of modern systems is to devise systems which are easy to use but extraordinary hard to break. For practical purposes it doesn't matter if an encryption can be broken after 10,000 years of effort!

The simplest digital encryption systems rely on digital random number generators like the shift register system described in the last chapter. The message sender and receiver arrange to use similar systems starting in the same state and use the string of bits their generators produce as electronic one time pads. In these systems the initial state of the random number generator represents a *Key* to deciphering the signal. To crack such a message a codebreaker needs to know both the encryption scheme and the value of the key. The indicator of the Enigma system was used to tell the receiver the key for that particular message.

Clearly, if a codebreaker knows the details of the encryption system, the signal can be broken as soon as the key is discovered. For this reason, a system using a shift register of modest length isn't very secure. The codebreaker can use the *Brute Force Attack* of trying every possible starting state (i.e. every possible key) until finding one which turns the encrypted signal into a sensible message. To avoid this, the encryption scheme should be capable of producing an enormous variety of patterns. (i.e. a random sequence which only repeats itself after a very long sequence — ideally infinite.) This would mean that trying every possible key would take far too long to be worth trying unless the codebreaker is really desperate! For this reason chaotic and semi-chaotic systems have attracted the interest of codemakers.

Various encryption schemes have been devised during recent years. Here we will use as an example a *Trapdoor* system of the general type outlined initially by Diffie and Hoffman (*New Directions in Cryptography*, 1976). The specific example we will examine is the Rivest, Shamir, Adelman (RSA) system. The name 'trapdoor' comes from the analogy that it is easier to fall down through an open trapdoor than to climb back up through it

203

again. In this case, given the appropriate information, the encryption method is designed to be easy in one direction and — without the right key — virtually impossible in the other. The basic requirements of a trapdoor system can be given in terms of four conditions which it must satisfy:

i) There should be a *Forward Algorithm, F*, for converting an input message, *X*, into its encoded form, *Y*, which uses some form of key, *K*, i.e.

$$Y = F\{K, X\} \qquad \qquad ...(21.1)$$

ii) This encrypted message can easily be decoded given another key, *L*, (related to *K*) via the appropriate *Inverse Algorithm, I*, i.e.

$$X = I\{L, Y\} \qquad \qquad ...(21.2)$$

iii) Without knowing the decoding key, *L*, it is computationally "impractical" to decode the message (i.e. it should take far too long to make it worthwhile).

iv) The number of possible key pairs, [*K,L*], is "very large" (i.e. it isn't worth trying each possible *L* in turn because it would take too long to discover the correct value).

The difficulty of breaking such an encoding scheme now depends upon how well it satisfies the third and forth properties. The terms, "impractical", and, "very large", in the above conditions have been placed in quotation marks because they are rather difficult to define precisely. This is because it isn't possible to devise a system which can be <u>guaranteed</u> to be invulnerable to attack. The power of computing systems, and the skills of codebreaking mathematicians, tend to increase with time. A code which was once considered unbreakable may soon fall victim to progress!

Any particular pair of functions or algorithms, *F* and *I*, are said to be *symmetric*. That is, *F* is the inverse of *I* and *I* is the inverse of *F*. Well-designed trapdoor systems have a number of interesting properties which can be very useful in practice. For example, it is possible to give someone the encoding algorithm, *F*, and key, *K*, without letting them know how to <u>decode</u> a message. This is because the function, *I*, and the value of the decode key, *L*, aren't (or at least, shouldn't be!) obvious from *F* and *K*.

We can now imagine a situation where various people have devised their own, individual, forward and reverse algorithms (equivalent to choosing a particular pair of [*K,L*] values), F_i and I_i. Each person can then freely publish the details of their particular forward algorithm (i.e., publish their key value K_i) with confidence that they remain the only one able to decode any messages sent to them using that algorithm! The published value of K_i is then called their *Public Key*. The corresponding inverse key value L_i which they keep secret is their *Private Key*.

One particularly interesting property of these systems is that it is possible to 'sign' messages. To see how this works, consider the situation where a sender, A, wants to communicate a secret message to a receiver, B, and also wants to give B proof that the message cannot have come from anyone but A.

Each of them chooses their own function/algorithm pair, F and I (i.e. selects their $[K,L]$ pair). They both then publish their forward algorithms. A takes his message, X, and initially encodes it into a new form, Y, using his own <u>inverse</u> function

$$Y = I_a\{X, L_a\} \qquad \qquad ...(21.3)$$

he then re-encodes this using the receiver's public forward algorithm into

$$Z = F_b\{Y, K_b\} = F_b\{I_a\{X, L_a\}, K_b\} \qquad ...(21.4)$$

and the message is then transmitted in the form of the signal, Z. Now, even if it is received by anyone else, only the intended recipient, B, has the inverse function, I_b, to recover

$$I_b\{Z, L_b\} = I_b\{F_b\{I_a\{X, L_a\}, K_b\}, L_b\} = I_a\{X, L_a\} = Y ...(21.5)$$

Having converted the message back into the form, Y, the receiver can now use A's public key to perform

$$F_a\{Y, K_a\} = F_a\{I_a\{X, L_a\}, K_a\} = X \qquad ...(21.6)$$

and obtain the original message. Given that only the sender possesses L_a, the message <u>must</u> be genuine if it can be de-encrypted using F_a. Similarly, the message is secure from eavesdroppers if only the intended receiver, B, possesses L_b. Hence it is possible to send a secure, signed, message even though K_a and K_b (and hence $F_a\{\}$ and $F_b\{\}$) are public knowledge.

In the RSA system the algorithms reply upon the properties of prime numbers. To design a particular coding scheme we must first choose a pair of 'large' prime numbers, P and Q, which we use to calculate the number

$$N = PQ \qquad \qquad ...(21.7)$$

Next, we select another 'large' integer, R, which is prime relative to (i.e neither is a factor of the other) the integer

$$S = (P - 1)(Q - 1) \qquad \qquad ...(21.8)$$

205

We should now find that there is just one number, E, in the range $1 \leqslant E \leqslant S$, which is such that

$$(ER) \bmod S = 1 \qquad \ldots (21.9)$$

having found this value we can publish the number, E, as the public key for our personal algorithm. R is the secret key kept for deciphering messages. (Note that the enciphering and deciphering processes must also agree on the value of N. This is also therefore 'public information'.)

The message sender and receiver now agree to represent each possible message symbol as an integer, x_i, in the range $1 < x_i < N$. For text messages each symbol can be a collection of two or more successive letters from the message. Each message symbol is then converted into an enciphered signal symbol

$$y_i = x_i^E \bmod N \qquad \ldots (21.10)$$

The received message can then be deciphered using

$$x_i = y_i^R \bmod N \qquad \ldots (21.11)$$

Here 21.10 and 21.11 are the forward and reverse algorithms. Clearly, these calculations are fairly quick and easy given a decent computer provided that you know the correct value of R or E. The problem for codebreakers is that, when P and Q are very large it can be 'very difficult' to discover the value of R from knowing N and E.

For this system to work well it is important to bring together the message symbols into groups large enough to use as many as possible of the available N integer values. For example, a system which just used the 256 ASCII codes for text wouldn't make effective use of a system where N $\approx$100 000 since only a few of the codes would be used. Instead, it makes sense to group pairs of letters of the message to make $256 \times 256 = 65{,}536$ message symbols — hence using over half the possible integers. The reason for this is that the RSA system we've described is just a very sophisticated substitution code. We must therefore ensure that N is much bigger than the length of the signal traffic. Otherwise the signals may be vulnerable to straightforward entropic attack.

Summary

You should now understand how simple *Substitution* encryption works and why it is vulnerable to *Entropic Attack*. That we can protect encrypted signals from attack be combining the message information with a string of randomly varying values. As a result, systems which employ 'true random numbers' cannot be deciphered without knowledge of the random number sequence used. You should now also understand that the practical problems of secretly transferring true random *One Time Pads* (or their equivalent) can be avoided by providing the transmitter and receiver with an agreed *Randomising Algorithm* which enables then to generate the same, apparently randomly varying, sequences for enciphering/deciphering information. However, it should be clear that these methods never produce true random sequences, hence it is possible to break the codes given enough information about the algorithm used.

Chapter 22

One bit more

22.1 Problems with many bits

In Chapters 9 to 11 we saw how CD systems work. One of the subjects this included was the use of *multi-bit* digital-to-analog convertors (DACs) to turn the digital values back into an analog waveform. For various practical reasons it's difficult to make accurate multi-bit DACs. Figure 22.1a illustrates a system which takes an input voltage, converts it to an *n*-bit digital value, and then turns that value back into an output voltage. This represents a sort of 'minimal' digital information communication system.

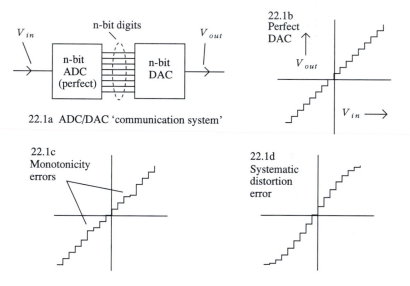

22.1a ADC/DAC 'communication system'

Figure 22.1   Signal transfer using 'back to back' ADC–DAC pair.

The ability of the system to *transfer* a signal from input to output can be represented by a graph showing how the output voltage, V_{out}, from the DAC varies with the input voltage, V_{in}, presented to the analog-to-digital convertor (ADC). A plot of this kind is called the *Signal Transfer Curve* of the system. For a perfect matching ADC/DAC pair we would expect the system to have a 'staircase' transfer curve of the kind illustrated in figure 22.1b. The steps of this staircase are produced by the quantisation produced when the ADC represents each input voltage as an appropriate digital value. For an ideal system all the steps should have the same heights and widths. The resulting staircase shape is said to be *Monotonic.*

Multi-bit DACs use a variety of techniques to convert digital values back into an analog voltage. Unless they're perfectly made, these produce two general sorts of errors. Figure 22.1c illustrates the effects of *Monotonicity Errors.* Here, imperfections in the DAC mean that some input digital values produce incorrect output voltages. The effect of this is to lift or lower some of the steps in the transfer staircase. Another type of problem is illustrated in 22.1d. This shows an overall or *Systematic* nonlinearity where imperfections in the DAC cause the output voltage to be wrong by an amount which varies relatively smoothly with the input. Both types of imperfection will cause the output signal to become distorted.

One way to avoid this distortion is, of course, to make and use very good ADC/DAC chips! This solution is OK for the music companies who can afford to spend thousands of pounds on ADCs. They can also afford to employ technicians to regularly check the ADCs and replace/adjust them if they aren't working properly. However, those of us who just want to buy and play CDs expect the player to work well without being outrageously expensive or requiring regular adjustment. The DAC systems in CD players should therefore be reliable, accurate, and (relatively) cheap.

As we saw in chapter 8, *Dither* can be used to suppress the effects of the basic staircase quantisation provided that it is applied <u>before</u> an original analog signal is digitally sampled. Dither can also be used to reduce the effects of monotonicity errors in either the ADC or the DAC. It does this by producing a result which is effectively 'averaged over' a number of levels, essentially smoothing over small-scale errors. The bad news is that dither can't totally remove monotonicity error distortions. The good news is that — unlike the dither used to suppress ADC quantisation effects — dither added to the <u>digitised</u> signal before digital to analog conversion can reduce monotonicity error problems. In a practical example like the CD audio system this means that dither has to be added during recording to remove quantisation effects, but the CD player can employ dither to reduce any tendency it has to distort the signals it recovers from the CD. This is one of the reasons why some manufacturers proudly boast that their CD players 'use dithering' to achieve improved performance.

In practice <u>all</u> multi-bit DACs will suffer from monotonicity errors at some level, however small. To avoid this problem altogether many modern CD

209

systems employ *one-bit* digital to analog conversion systems. This chapter will examine how these work as they make an excellent example of how information can be converted from one form into another without loss.

22.2 One bit at a time

A conventional 16-bit DAC can output 2^{16} different output voltages, each corresponding to a different input digital value. A one-bit DAC can only produce 2 possible output voltages — 'high' or 'low'. Figure 22.2 illustrates how a typical one-bit system works.

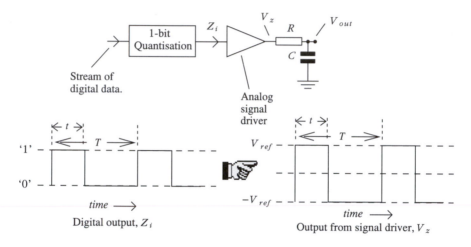

Figure 22.2   Basic 'one-bit' DAC system.

The system uses one digital line, Z_i, which will either be 'high' (='1') or 'low' (='0') at any instant. This digital level is used to operate an *Analog Signal Driver* — an amplifier whose output is $+V_{ref}$ when Z_i = '1' and $-V_{ref}$ when Z_i = '0'. The choice of the value V_{ref} isn't important provided it is constant whilst the circuit operates.

The digital output consists of a stream of pulse 'cycles' with a period or *interval, T*. The duration or *width* of each pulse is t. The output from the driver is passed through a time constant (or some other low-pass filter) chosen so that $RC \geqslant T$. We can think of the input waveform, V_z, as being a combination of a d.c. level plus some a.c. components which give the pulse shape. The output voltage, V_{out}, at any moment will therefore roughly equal the value of V_z averaged over the most recent pulse cycles.

210

Now the average value of V_z during one pulse cycle will be

$$\langle V_z \rangle = \frac{t\,V_{ref} - (T - t)\,V_{ref}}{T} \qquad ...(22.1)$$

where the angle brackets indicate that we're talking about an averaged or smoothed value. We can therefore expect that

$$V_{out} \approx \frac{t\,V_{ref} - (T - t)\,V_{ref}}{T} \qquad ...(22.2)$$

which can be simplified into the form

$$V_{out} \approx \frac{V_{ref}\,(2t - T)}{T} \qquad ...(22.3)$$

As a result — provided the value of RC is large enough — the filter will smooth out the pulses and the output will be a d.c. level whose value depends upon the ratio $(2t - T)/T$. Since t can't be less than zero or greater than T this means we can obtain any output voltage we wish in the range $-V_{ref} \leqslant V_{out} \leqslant V_{ref}$ by choosing appropriate values for the pulse width, t, and the cycle period, T.

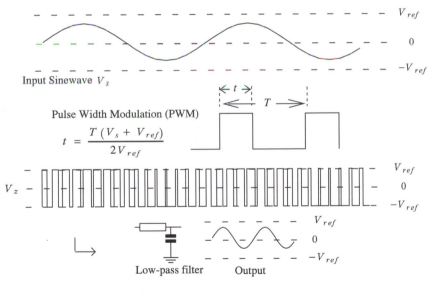

Figure 22.3   Pulse width modulation.

Two basic methods which use this technique are *Pulse Width Modulation* (PWM) and *Pulse Density Modulation* (PDM). Figure 22.3 illustrates the use of PWM to convey information which can be converted back into an analog sinewave. Here the pulse cycle period, T, is kept constant and the pulse width is varied according to the signal voltage level, V_s, we wish the system to output. Since we require the final output level to equal V_s we can rearrange equation 22.3 and say that

$$t = \frac{T(V_s + V_{ref})}{2V_{ref}}$$ ... (22.4)

PDM keeps the pulse width fixed and alters the cycle period to achieve the required smoothed output voltage level. In this system we therefore require

$$T = \frac{2V_{ref}t}{(V_s + V_{ref})}$$ ... (22.5)

This produces the kind of signals illustrated in figure 22.4

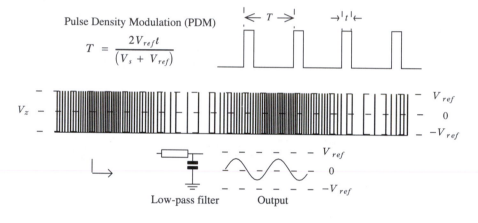

Figure 22.4 Pulse density modulation.

Both PWM and PDM are used in data telemetry/communications systems to send information about analog levels in the form of a 'digital' signal — i.e. one whose level is either '1' or '0' at any instant. These forms of signal and the circuit which converts them back into analog form are called 'one-bit' since only two possible levels are involved. In practice PDM has a disadvantage that the pulse interval becomes very long when $V_s \rightarrow -V_{ref}$. This means that the a.c. components of the signal then extend to low frequencies and may not be very well suppressed by the output filter. Hence PWM is usually preferred.

Perhaps the simplest way to make a CD player one-bit DAC is the method illustrated in figure 22.5. Here X_i represents the series of 16-bit values recovered from the CD (for simplicity we'll only consider one channel).

The subscript, i, indicates which sample in the series we're considering. So the samples appear in the regular sequence, $...X_{(i-2)}$, $X_{(i-1)}$, X_i, $X_{(i+1)}$, $X_{(i+2)}$, etc ... CD sample values are integers in the range $+2^{15}$ to $(-2^{15}+1)$. For this particular form of one-bit DAC to work we have to begin by adding 2^{15} to each sample value to produce the 'shifted' values X_i', which are all in the range 0 to $+2^{16}$. The shifted 16-bit sample values, X_i', are then loaded into a *Down Counter*. This counter is driven by a clock whose period is $t_c = T/2^{16}$, where T is the time interval between successive samples. Each clock pulse or 'tick' makes the counter reduce it's stored value by one. The counter provides a one-bit output, Z_i, which is '1' whenever the value stored by the counter is greater than zero. When the steady countdown reaches zero, Z_i is switched to '0' and this halts the counting process. As a result, each input sampled value loaded into the counter produces an output '1' pulse whose width $t = X_i' t_c$. Higher (more positive) signal levels produce wider pulses and lower (more negative) signal levels produce narrower pulses. The binary signal, Z_i, therefore represents the input stream of 16-bit values converted into one-bit PWM form. Hence V_{out} should be the required analog music signal.

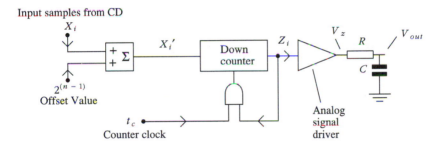

Input samples from CD

Figure 22.5 Simple PWM one-bit DAC system.

In practice this type of one-bit DAC would be difficult to make work well. The main reason for this is that, for CDs, the sampling interval $T = 22\cdot675$ μs. This means that the countdown clock required would have a clock interval of $t_c = T/2^{16} = 0\cdot346$ ns. This means a clock frequency $f_c = 1/t_c$ of 2·89 GHz! Clock oscillators and digital logic operating at this high frequency are currently too expensive for use in consumer products (although this may, of course, change in the future). As a result this method isn't suitable for normal domestic systems. Another problem is that we would require an excellent output filter to pass the required audio information whilst blocking the high frequency pulse shapes. This difficulty could be eased by *Oversampling* to increase the input sample rate and hence reduce T. However this would also increase f_c, making the

213

system even less practical for normal applications.

22.3 From many to one

To overcome the problems described above we need to make use of the oversampling and *Noise Shaping* techniques we met in an earlier chapter. Most CD player manufacturer like to devise their own way to perform this operation — and, naturally, their newest way is always the 'best'. For this chapter we will take one method, the use of a *Delta-Sigma* (or '$\Delta\Sigma$') DAC. Before looking at this in detail it's worth making a general point about systems which oversample and change the 'number of bits per value' of a stream of digital data.

In previous chapters we saw how the first generation of Philips DACs used 4× oversampling to permit the use of a 14-bit DAC to recover all the input 16-bit information. In general, we can describe a *p*× oversampling and noise shaping system as taking in m samples per second and generating $m' = pm$ output samples per second. Each input sample will have n bits and each output value will have n' bits. The rate at which bits of information enter the oversampler will therefore be mn. The rate at which they emerge will be $m'n'$. From the basic arguments of information theory we can expect that — provided the system works in a sensible way — no information need be lost provided that $m'n' \geqslant mn$. This is because the amount of information conveyed in a given time depends upon the rate of bit transfer. The effect of having fewer bits per sample can be counteracted by having more samples. On the basis of this argument we can expect that a system which oversamples by at least 16× should be capable of providing a stream of 1-bit output values which carry all the information from an input stream of 16-bit samples. In practice — as we might expect — it is normally advisable to ensure an *Oversampling Ratio* which ensures that $m'n'$ is significantly greater than mn to avoid effects of any imperfections in the signal conversion process.

22.4 First order delta–sigma conversion

Figure 22.6 illustrates a *First Order $\Delta\Sigma$ Convertor*. Here X_i represents the initial input stream of 16-bit integer values coming from the CD. These are passed through a *Transverse Digital Filter* similar to that described in an earlier chapter. This produces an oversampled stream of values, X_i' which is presented to a Δ (or 'difference') unit. This compares the current oversampled value with the single output bit, Z_i, currently emerging from

214

the convertor. For this part of the circuit Z_i is treated as the 'most significant bit' of a binary number having the same number of bits as each oversampled value — i.e. for, say, 4-bit values, Z_i 'high' would be treated as $+2^3$ and 'low' as $(-2^3 + 1)$. The Δ unit subtracts the value of Z_i from X_i' and passes the result, Y_i, on to a *Digital Integrator*.

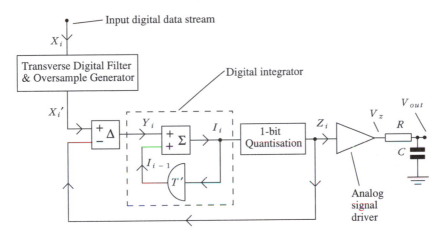

Figure 22.6 First order delta–sigma DAC system.

The integrator consists of a Σ ('sum') unit and a delay gate. The delay feeds the output from the Σ unit back to its input, but delays it for one oversample clock interval. As a result, the output from the digital integrator $I_i = Y_i + I_{(i-1)}$, i.e. the sum of the current input to the Σ unit and the 'previous' output value. A steady input to this arrangement would cause the output to change steadily at a rate proportional to the size of the input, hence the combination of the Σ unit and the delayed feedback behaves as an integrator.

The output from the integrator is passed to a unit which simply tests whether the result is greater than zero or not. If $I_i > 0$ then it sets $Z_i = 1$. If not, it sets $Z_i = 0$. This output value is then used to control the output driver and fed back to the input Δ unit. The behaviour of this system is illustrated in figure 22.7. Note that the conversion process shown has deliberately been done too poorly for ideal conversion of 16-bit input values. This is to help make clear the characteristics of this form of DAC. The example only uses 8× oversampling whereas <u>at least</u> 16× would be required for no information to be lost. Also, the output filter does not reduce the effects of driver pulses very much. A better output filter and a

215

higher oversampling ratio would produce a much more accurate output analog sinewave.

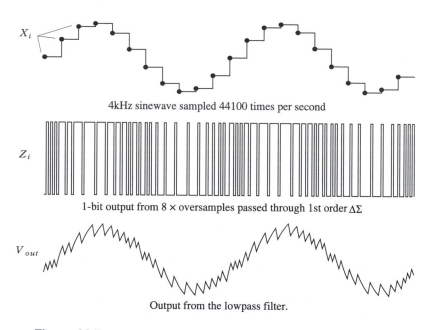

Figure 22.7   8 × oversampling first order delta–sigma DAC.

Examining figure 22.7 we can see that the output has the correct form, but has a high frequency 'frizz' error pattern superimposed upon it. Note that this error pattern essentially consists of frequencies between 4× to 8× the basic sampling rate and above. Hence this unwanted addition to the signal is at frequencies well above the audio range and could largely be removed by a better output filter. A higher oversampling ratio would increase the frequency of this error pattern and reduce its amplitude, making it easier to filter the unwanted 'frizz' off the wanted audio waveform. Current generations of *Bitstream* DACs (Philip's name for their one-bit systems) use 256× oversampling. As a result, the output error patterns they produce are mainly at frequencies around 256 × 44·1 kHz = 11·289 MHz and above. These frequencies are far enough from the audio band to be removable with relatively simple low-pass filters.

22.5 One last bit of chaos!

The first order $\Delta\Sigma$ considered above is the simplest member of a family of delta–sigma convertors. Figure 22.8 illustrates a second order $\Delta\Sigma$ DAC.

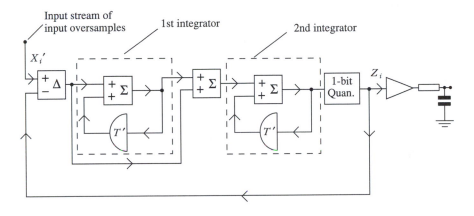

Figure 22.8   Second order delta–sigma DAC system.

Second order $\Delta\Sigma$ convertors are generally preferred to first order systems. This is mainly because their *Idler Pattern* behaviour is better.

The idler pattern is the output pattern of '1's and '0's the DAC generates to produce a steady analog output level. Figure 22.9 illustrates some idler patterns produced by a first order $\Delta\Sigma$ DAC which has a 6-bit range. (That is, the maximum +ve output corresponds to $X_i = 2^5$.) The top line shows the output pattern when the system is switched on and presented with a series of $X_i = 0$ values. The result is a series of alternating '1's and '0's. The lines below show the effect of increasing the input series of number to +4, then back to 0, then to −6, then to 0 once more.

In each case the output series will, when averaged over a reasonable time, give the correct value. However, this illustration shows two effects. Firstly, that an output of zero makes the DAC generate a repetitive squarewave sequence, '...10101010101...'. Secondly, that this sequence reappears whenever the input signal returns to 0. Note that values of X_i near to zero also produce waves which spend a large fraction of the time behaving like a '...10101...' squarewave. These patterns tend to concentrate their high frequency energy into a few strong frequency components. As a result we require a good low-pass filter to suppress them to the −100 dB level

217

required to ensure the full dynamic range of a CD system. In addition, the highly coherent, repetitive nature of these patterns means that, if they are 'mixed' or combined with any other repetitive signal, a low frequency 'whistle' may be produced in the audio output. Unwanted whistles of this kind can appear as a result of nonlinear beating with clock harmonics or with rf interference from other equipment.

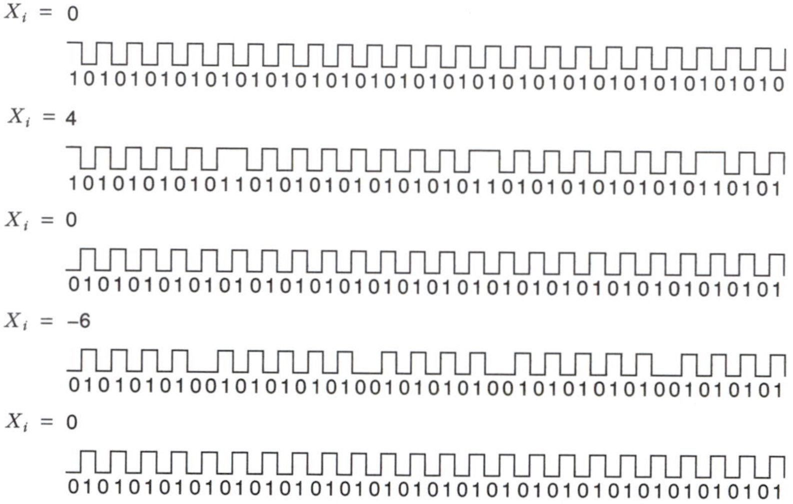

$X_i = 0$

1010101010101010101010101010101010101010101010101010

$X_i = 4$

1010101010110101010101010101101010101010101010110101

$X_i = 0$

0101010101010101010101010101010101010101010101010101

$X_i = -6$

0101010100101010101001010101010010101010100101010101

$X_i = 0$

0101010101010101010101010101010101010101010101010101

Figure 22.9 Output 'idler' patterns from a first order delta–sigma DAC.

To avoid these effects we require a DAC system which produces idler patterns which are more variable. Figure 22.10 illustrates some idler patterns produced by a second order $\Delta\Sigma$ DAC.

Note that in this illustration all the patterns shown are produced with an input series of $X_i = 0$ values. The top line of the illustration shows the pattern produced when the system is switched on. Each of the lines below it were produced after the X_i values had been moved away from 0 for a while and then returned to 0. From this illustration we can see that the second order system can generate a wide variety of idler patterns when required to produce a steady analog output of zero. In principle, any pattern of '1's and '0's can be produced provided that on average the numbers of each in a long time are equal. The pattern the system settles on depends on it's 'recent history' (i.e. the values stored in its integrators) when the input series returns to zero.

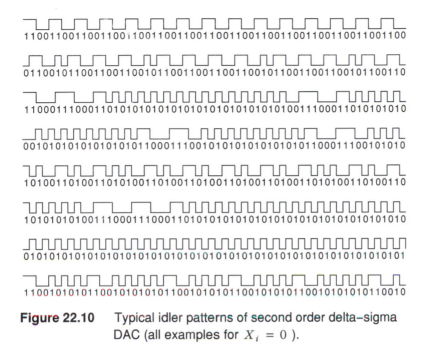

Figure 22.10   Typical idler patterns of second order delta–sigma
DAC (all examples for $X_i = 0$).

This variability of the idler pattern also applies to other, non-zero, series of X_i values. It means that the second order system tends to produce an idler pattern which varies from time to time. As a result, the unwanted high frequency spectrum of the idler pattern tends to consist of a large number of low-power frequency components and this spectrum changes from moment to moment as a signal is recovered. This tends to 'blur out' any unwanted mixing problems and hence the effect is to slightly alter the <u>noise</u> background of the audio signal. This effect is undesirable, but not as objectionable as unwanted whistling noises!

There is a measure of 'unpredictability' in which idler pattern a second order system will produce since the pattern depends upon the system's recent history. Third (and higher) order $\Delta\Sigma$ systems can also be made. These take this variability further and tend to produce *Chaotic* idler patterns. In itself, this behaviour is desirable since it ensures that idler-produced whistles become impossible. Unfortunately, high order systems sometimes become so unstable that the values stored in their integrators rise until they overflow the register sizes. This can produce disastrous signal distortion, hence higher order systems must be used with great care. (Fortunately, methods do exist to 'stabilise' them and avoid this effect.)

219

Note that the second order system does sometimes produce a regular idler pattern (a squarewave). To prevent this being a problem, commercial systems usually deliberately add a dither pattern to the the the X_i values fed to the DAC. This tends to 'break up' any repetitive behaviour — in effect it produces a DAC whose order is 'two-and-a-half'! The idler pattern this produces is semi-chaotic. The system remains stable and should not produce integrator overflows.

Summary

You should now know that *Multi-Bit* DACs can suffer from practical problems of *Non-Monotonicity* and *Systematic Errors* which can distort the output waveform. That these problems can be avoided by using a *One-Bit* DAC system. You should now also understand that any digital system which takes an input of m samples/sec, each n bits long, and outputs m' samples/sec, each n' bits long can ensure that no information is lost provided that $m'n' \geqslant mn$.

You should now understand how *Pulse Width Modulation* (PWM) and *Pulse Density Modulation* (PDM) can be used to generate a wave which can be averaged to obtain a required analog signal. That this principle of averaging over a pattern of '1's and '0's can be used to recover an analog signal from other one-bit systems. You should also now understand that simple PWM/PDM systems aren't currently suitable for CD players. That $\Delta\Sigma$ DACs can work well using a combination of noise shaping by *Digital Integrators*, and oversampling. That second order $\Delta\Sigma$ systems are preferred because they have less regular *Idler Patterns*, and that higher order $\Delta\Sigma$ systems can behave in a chaotic, unstable pattern.

Chapter 23

What have we here?

23.1 Distinguishing messages

In many of the previous chapters of this book we have been concerned with the need to be able to distinguish messages from random noise. We have also spent time looking at assessing the information content of messages and the information capacity of channels. One important topic, however, we have taken for granted. Up until now we have assumed that we can easily tell one message from another. We have also tended to take for granted that the chosen patterns or symbols in use can be easily distinguished from one another.

What is it that makes it possible to distinguish one message from another, and how can we choose a system that maximises our ability to recognise the meaning of signal patterns when they arrive?

Pattern library

Typical message pattern

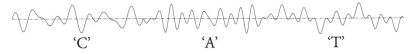

Figure 23.1   Patterns for sending messages.

In human terms, we tend simply to look at patterns and 'recognise' them. In effect this means we have a sort of mental library of patterns which we associate with given meanings. Figure 23.1 gives an illustration of this process. Here each letter of the alphabet is represented as a specific

221

pattern. In this case we can imagine each pattern as showing how a received quantity – e.g. a voltage – varies with time. We then compare a message when it is presented to us and choose the patterns in our memory which seem 'closest' to what has been presented to us. In this case, when we look at the message pattern shown an an example in figure 23.1 we can compare it with the library of patterns for the letters and recognise the message as being 'CAT'.

This is all very well when we are happy to recognise message patterns or symbol shapes by eye. More generally, however, we need to be able to define, mathematically, what sort of process takes place when we are identifying and recognising patterns. There are number of reasons for this. For example, we may get bored and want to automate the message recognition process. In practice we may want the messages to control equipment, or to send data at high rates, etc. In addition we may find that the received messages aren't always easy to recognise.

Figure 23.2   A message altered by noise.

Figure 23.2 illustrates this situation. Here the pattern for just one letter has arrived. Which one is it? If we compare, by eye, figure 23.2 with the library shown in figure 23.1 we can decide that it is most probably a 'B', but it isn't easy to tell, and the presence of the added noise may mean that we have made an error. In all real signal transfer systems we can expect some random noise to be present, so we need some objective, mathematical, way to take the signals as they arrive and compare them with our library. This can then serve two primary purposes. Firstly, to speed up and automate the process. Secondly, to help us assess how likely it is that we have identified the right message.

There is also a further useful advantage to employing a mathematical approach as it then helps us to decide how we can choose the 'best' library of patterns to aid the recognition process. When recognising them by eye, the best set of library patterns would be chosen so that they all looked as different to one another as possible. We will see later what 'best' means here in a more objective mathematical sense.

23.2 Correlation

The standard mathematical technique that is employed to compare patterns and assess how 'similar' they are is called *Correlation*. Given a library of continuous functions, $L_i\{t\}$, (i.e. a set of functions; $L_A\{t\}$ to define the pattern for 'A', $L_B\{t\}$ to define the pattern for 'B', etc.), we can define the correlation of each of these with some input, $x\{t\}$, using the integral

$$C_i \equiv \frac{1}{\alpha_i} \int_0^T x\{t\} L_i\{t\} \, dt \qquad \ldots (23.1)$$

By performing a series of these integrals for each $L_i\{t\}$ in our library we can obtain a set of C_i values which we can then use to help us decide which – if any – of the library patterns is most similar to $x\{t\}$. The term, α, is a *Normalisation* factor whose value may be defined using the expression

$$\alpha_i \equiv \sqrt{\left[\int_0^T x^2\{t\} \, dt \right] \times \left[\int_0^T L_i\{t\} \, dt \right]} \qquad \ldots (23.2)$$

This ensures that any values we obtain are always in the range

$$-1 \leqslant C_i \leqslant +1 \qquad \ldots (23.3)$$

Normalisation is convenient as it means we can concentrate on similarities between the shapes or patterns without having to worry too much about their amplitudes or lengths.

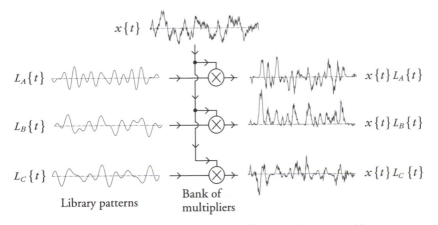

Figure 23.3 Multiply together library patterns and input.

To illustrate why the correlation is useful we can examine figure 23.3. This shows the results we get when we try multiplying the input pattern by each of the library patterns. Looking at each of the results we can see that the pattern for $x\{t\}L_B\{t\}$ is distinctive in that the result is generally positive (or zero). The patterns obtained by multiplying the input by other library shapes all tend to fluctuate in both the positive and negative directions with no obvious preference.

This tendency for one result to be positive is quite significant. It indicates that the two patterns being multiplied together tend to share the same sign at every instant. It also means that the integral of $x\{t\}L_B\{t\}$ tends to give a distinct positive result. Integrating the other output patterns, for $x\{t\}L_A\{t\}$, etc., tends to give values of much smaller magnitude which are just as likely to be negative as positive. This difference in behaviour makes the result of integrating $x\{t\}L_B\{t\}$ stand out from the crowd, indicating that we may have identified a special relationship between the noisy input $x\{t\}$ and the pattern for 'B', $L_B\{t\}$. Hence we can use it as a way of recognising that in this case the pattern, $x\{t\}$, is probably that which signals a 'B' despite the noise which disguises this fact.

In practice, we would often sample the signals and patterns, and then use numerical summations rather than integrals. Theoretically, this gives the same results as if we had used (as indicated in figure 23.3) analogue multipliers on the waveforms, but given the power of digital computers this sample-based method is usually more convenient in practice. It also makes the argument for what is happening slightly clearer, so we can adopt the approach here.

When dealing with series of sampled values we can define the correlation between a pair of patterns in terms of a series

$$C\{x, y\} \equiv \frac{1}{\alpha_{xy}} \sum_{j=1}^{N} x_j y_j \Delta t \qquad \text{... (23.4)}$$

where the information in one pattern is now represented by the series of values $x_1, x_2, \ldots x_j, \ldots x_N$ each of which records the level of the pattern at a time $j\Delta t$ from the start of the pattern. In a similar way, the other pattern's information is represented by the values $y_1, y_2, \ldots, y_j, \ldots, y_N$ taken at the same moments. As we have seen in earlier chapters, these series can hold all the information about the original patterns provided that the sampling interval, Δt, is small enough to ensure that we have satisfied the sampling theorem.

In terms of these sets of sampled values the normalisation factor will be

$$a_{xy} \equiv \sqrt{\sum_{j=1}^{N} x_j^2 \, \Delta t \times \sum_{j=1}^{N} y_j^2 \, \Delta t} \qquad \text{... (23.5)}$$

Before considering the specific patterns in the example illustrated earlier it is useful to understand the general properties of this numerical correlation. Once these are understood the results which arise when we use correlation for pattern recognition should become clear.

Combining expressions 23.4 and 23.5 we can say that

$$C\{x, y\} \equiv \frac{\langle x_j y_j \rangle}{\sqrt{\langle x_j^2 \rangle} \times \sqrt{\langle y_j^2 \rangle}} \qquad \text{... (23.6)}$$

where the angle brackets are used to indicate a quantity averaged over the N values of each summation. The actual value of N, and that of Δt have vanished from expression 23.6 as the were present in both denominator and numerator and hence have been cancelled out and removed.

The above expression is simpler than those given earlier, so it makes the properties of the correlation clearer. For example, we can see that the normalisation term is essentially the product of the rms levels of the two series. The effect of the normalisation is therefore to produce the same result as would occur if the input patterns happened to have rms levels of $\langle x_j^2 \rangle = \langle y_j^2 \rangle = 1$. To further simplify the argument we can therefore assume that the levels <u>have</u> been prearranged so that this is the case.

The precise result we obtain will obviously depend upon the details of the sets of values, x_j and y_j. We can however obtain some general conclusions based upon assuming that the patterns we have sampled are intended to *efficiently* communicate information. From previous chapters we already know that an efficient signal has statistical properties similar to those of random noise. The above normalisation implies that the most typical level of $|x_j|$ and $|y_j|$ will also be unity since they are the square root of unity. Hence the most likely value, statistically, of $|x_j y_j|$ will also be around unity.

When calculating the average value of $x_j y_j$ we now say that there will be N contributions, each having a typical magnitude of unity, but with an actual magnitude and sign which varies randomly from sample to sample. As a result, when the patterns are unrelated, the most probable value of the sum of this product will statistically be

225

$$\sum_{j=1}^{N} x_j y_j \approx \sqrt{N} \qquad \text{... (23.7)}$$

in effect, this is an example of random, incoherent, addition, so the level only tends to grow as the square root of the number of contributions. When we divide the sum by the number of contributions to obtain the mean level we therefore obtain a most probable result of

$$C\{x, y\} \approx \frac{\sqrt{N}}{N} = \frac{1}{\sqrt{N}} \qquad \text{... (23.8)}$$

Note that this only indicates the most probable magnitude of the result. The actual value can vary, and is just as likely to be negative as it is positive. When correlating two unrelated patterns whose amplitudes have already been normalised to $\langle x_j^2 \rangle = \langle y_j^2 \rangle = 1$ it is probably better to regard the result as being

$$C\{x, y\} \approx \pm\frac{1}{\sqrt{N}} \qquad \text{... (23.9)}$$

to make it clear that result is really the most likely size for a range of possibilities. Pairs of patterns which produce results like this are said to be *Uncorrelated* as the calculation indicates that they are unrelated.

23.3 The effects of noise

Now consider what happens where there is a relationship between the input pattern, x_j, and the pattern, y_j, which we are correlating it with. For clarity we can now define the input to be

$$x_j \equiv \beta y_j + n_j \qquad \text{... (23.10)}$$

i.e. the input has two components, one being a scaled version, βy_j of the signal pattern we are now comparing it with, the other being random noise, represented by the series of value, n_j. As before, for the sake of simplicity we will assume that we have scaled the patterns so that they are normalised to $\langle x_j^2 \rangle = \langle y_j^2 \rangle = 1$. The correlation now has the form

$$C\{x, y\} = \frac{1}{N}\left[\sum_{j=1}^{N} \beta y_j^2 + n_j y_j\right] \qquad \text{... (23.11)}$$

The first term represents the part of the input, x_j, that has the same pattern as the series, y_j, we are correlating it with. Hence it produces a value which is just equal to β. The above is therefore equivalent to

226

$$C\{x, y\} = \beta + \frac{1}{N} \sum_{j=1}^{N} n_j y_j \qquad \qquad \text{... (23.12)}$$

The second term is a summation of the series of $n_j y_j$ values.

Since we can expect the random noise pattern, n_j, to have no relationship with the pattern of, y_j, this is similar to the result given in expression 23.7. However in that case we were correlating two patterns whose sizes had already been scaled to ensure that $\langle x_j^2 \rangle = \langle y_j^2 \rangle = 1$. We have not scaled the size of the series, n_j, in this way. In this case, therefore, we find that

$$\frac{1}{N} \sum_{j=1}^{N} n_j y_j \approx \pm \frac{\sqrt{\langle n_j^2 \rangle \times \langle y_j^2 \rangle}}{\sqrt{N}} \qquad \qquad \text{... (23.13)}$$

This is really just a more general form of the result used earlier. As before, we have pre-set $\langle y_j^2 \rangle = 1$, so we can say that the above means that

$$C\{x, y\} = \beta \pm \frac{\sqrt{\langle n_j^2 \rangle}}{\sqrt{N}} \qquad \qquad \text{... (23.14)}$$

In the absence of any noise contribution we would expect $\beta \rightarrow 1$ and all $n_j \rightarrow 0$. The result would be a correlation value of unity and we would say that the x_j and y_j patterns were perfectly *Correlated*. We would then use this as evidence to unambiguously say that the patterns x_j and y_j were the same. The presence of the noise alters this in two ways. To understand these it is useful to notice that expressions like $\langle x_j^2 \rangle$ essentially indicate the 'mean power' of the patterns.

The first consequence of the presence of the noise is that the signal pattern, βy_j, now only provides a fraction of the total power of x_j. It actually provides just β^2 of the total. Since we have scaled the size of the x_j values to get unity total power from the input combination of signal plus noise, we can say that

$$\langle x_j^2 \rangle = \beta^2 \langle y_j^2 \rangle + \langle n_j^2 \rangle = 1 \qquad \qquad \text{... (23.15)}$$

Since we also know that we have arranged for $\langle y_j^2 \rangle = 1$ and we have a non-zero noise power it follows that β will have a value less than unity, and that the greater the noise level, the smaller β will be. The primary effect of the presence of the noise is therefore to reduce the typical level we get when we correlate a signal with the 'correct' pattern that it contains since we are seeking a correlation value approaching unity as a sign that we have identified the signal pattern.

The second consequence of the presence of the noise can be seen by looking again at expression 23.14. The second term in that expression indicates that there will be a level of uncertainty or error in the value obtained by performing the correlation. This is just the usual, inevitable, result we would expect from the basic ideas of Information Theory. i.e. having made a 'measurement' (in this case a test to see if a specific signal pattern is present) we can expect some level of uncertainty in the result due to the presence of noise. This limits the amount of information we can gather, in this case meaning we can never be 100% certain we have correctly identified the signal pattern.

To assess this level of uncertainty we can make use of the value of β to link the input and output Signal to Noise Ratios of the measurement process implied when we correlate the input against the 'correct' pattern.

From expression 23.15 it follows that the relative level of the noise must be such that $\langle n_j^2 \rangle = 1 - \beta^2$. This allows us to link the input Signal to Noise Ratio (SNR) of x_j (i.e. the power ratio of the actual signal level, y_j to the noise power level of n_j) to the value of β via the expression

$$\text{SNR}_{in} \equiv \frac{\beta^2}{1 - \beta_2} \qquad \ldots (23.16)$$

Looking back at expression 23.14 we can take the two terms and identify the first with the detected signal level, and the second with the output noise level. Since Signal to Noise Ratios are always power or energy ratios we have to square these to obtain an output result of

$$\text{SNR}_{out} \equiv \frac{N\beta^2}{1 - \beta^2} = \text{SNR}_{in} \times N \qquad \ldots (23.17)$$

This result tells us that the correlation process enhances the SNR and this can help us to 'pick out a signal from noise'.

The value of the ratio $\text{SNR}_{out} / \text{SNR}_{in}$ is often called the *Process Gain*. The longer the pattern sequence, the higher the process gain, and the greater the improvement we can obtain. As a result we can often begin with a situation where the 'raw' or input SNR is less than unity (i.e. the input signal power is less than the input noise power) and by performing a correlation with the relevant pattern obtain a clear detection of the presence of the signal pattern with a final SNR well above unity. Correlation is therefore a very valuable technique when we are seeking patterns which may be submerged in noise, as well as when we want to reliably decide which pattern from a possible set has arrived.

This improvement should not really be a surprise as it is very similar to results obtained in many cases described in earlier chapters. The accuracy or confidence of the output rises with the length of the sequences of values we have available. In fact Correlation is a process we have already met in this book in various disguises. For example, the integration technique used for signal averaging in Chapter 15 essentially correlates the input with the 'pattern' of a steady level. Similarly, the Phase Sensitive Detection process described in Chapter 16 is a way of correlating an input with a square-wave pattern of a specific frequency and phase.

Applying the arguments of the Sampling and Nyquist Theorems we can expect that a series of values, x_j, will form a complete record of a continuous pattern provided they represent the values taken at instants less than $\Delta t = \frac{1}{2B}$ apart. In this situation a series of N samples will take up a record duration of $T = \frac{N}{2B}$. This means that we can say that when considering continuous signal patterns which arrive with superimposed noise, the value of the Process Gain will be equal to

$$G_P = N = \frac{T}{\Delta t} = 2BT \qquad \ldots (23.18)$$

When we wish to estimate the amount of SNR enhancement a correlation process will provide we can now choose to use N for situations where we have sets of sampled values, or the signal's bandwidth and duration when dealing with continuous waveforms.

23.4 Signal recognition using correlation

Since Correlation can provide Process Gain we can use it to detect signals in the presence of a noise power level which may be higher than the signal's power level. This often means receiving a noise-dominated input and searching it for one or more 'known' patterns which are expected. This raises two issues which we have to resolve in order to be able to recognise signals. Firstly, how can we optimise our chances of being to tell which signal has arrived? Secondly, how can we tell when a signal arrives when it is buried in noise? Let's start with the first question.

It should be fairly clear from the previous sections of this chapter that we can hope to identify which pattern has arrived by performing correlations and finding which of our library gives the largest correlation value. To be able to do this as effectively as possible we'd like to arrange for two things to be true. Fairly obviously, we'd like to arrange for the highest possible input SNR to make the signal stand out from the noise. This being the case, we can hope to minimise the effect of the noise on the correlated output when we find the right message pattern and get a correlation value that approaches unity. However, in addition to this, we will also find it useful to make the actual signal patterns as 'different from each other' as possible. To understand what this means, we can assume that the actual noise level is small enough to be ignored, and that the input signal is a specific choice, $L_k\{t\}$, from our library, $L_j\{t\}$. To maximise detectability we want to arrange – if we can – that

$$\frac{1}{a_i} \int_0^T L_k\{t\} L_i\{t\} \, dt \equiv 0 \qquad \text{when} \qquad k \neq i$$

$$\frac{1}{a_i} \int_0^T L_k\{t\} L_i\{t\} \, dt \equiv 1 \qquad \text{when} \qquad k = i \qquad \ldots (23.19)$$

This means that, in the absence of any noise, we should find that only one correlation will give a value of unity, when we happen to try the library pattern for which $k = i$. All other correlations will give a result of zero. This makes the 'correct match' stand out as clearly as possible.

Sets of functions or series of values which satisfy expression 23.19 are said to be *Orthogonal* functions or series over the interval of the integral or summation. Choosing such an *Orthogonal Set* is therefore desirable when selecting the signal patterns we are using as they optimise our ability to distinguish one message from another. In fact, we have already seen this behaviour in earlier chapters as sinewaves and cosinewaves are used as the Orthogonal functions which form the basis for Fourier Transformation.

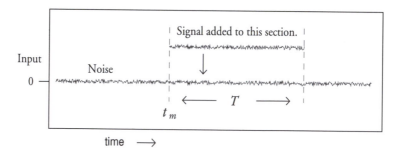

Figure 23.4   Signal hidden in a section of noise.

By choosing a suitable set of functions we can minimise the risk of one message being mistaken for another simply due to their being 'similar'. However, we still have to deal with the uncertainties introduced by noise, and may often have the problem of recognising the time when a signal has arrived if it is weak compared to the noise level. To see how we can deal with this situation we can use the example illustrated in Figure 23.4. This shows a signal pattern, of duration, T, added into a random noise pattern at a location starting at time, t_m. If the signal power is low compared to that of the noise, the signal seems to 'vanish' when we just look at the raw combination. In addition, as an efficient signal, the signal's pattern can be expected to have statistical properties similar to those of the noise, so it has no obvious features that show when it occurs.

The situation shown in figure 23.4 represents the situation which arises when we are monitoring a communication channel, waiting for the arrival of a signal, but not knowing when it might arrive. As before, we can consider the situation in terms of series of sampled values, but whilst doing so bear in mind that similar arguments and conclusions will arise for continuous functions and patterns.

We can now represent the input as a series of values, v_i, which arrive at instants, $j\Delta t$. The signal pattern we are looking for (which may be just one of many we look for in parallel) can be represented as a series of N values, $y_1, y_2, \ldots, y_j \ldots, y_N$. Once more than N input values have arrived we can collect a consecutive series of them, starting at an instant $p\Delta t$, and correlate this with the pattern we are looking for. For clarity we can ignore the normalisation terms and just say that the result will be a correlation value

$$C\{N, p\} \propto \frac{1}{N} \sum_{j=1}^{N} \left(y_j v_{j+p}\right) \Delta t \qquad \ldots (23.20)$$

As new sample points (new input data) arrives we can repeatedly recalculate this value for new values of p and correlate later and later sections of the input against the *Key* pattern(s) in our library of possible signals, searching for a match that tells us which pattern has arrived, and its time of arrival.

Now the noise won't have any specific relationship with the patterns we are seeking, so, provided that N is reasonably large, this usually won't contribute a significant amount to the result of the summation. Similarly, if the signal pattern is efficient, each of its values will be *independent* in information theory terms from its companions. (This was discussed perviously at the start of Chapter 8.) This has an important consequence which we can understand by considering expression 23.21

$$C\{m\} = \frac{1}{N}\sum_{j=1}^{N} a_j a_{j+m} \qquad \text{...(23.21)}$$

Here $C\{m\}$ represents the result of correlating the pattern a_j with <u>itself</u>, but with an <u>offset</u>, m. We know from earlier in this chapter that when $m = 0$ the correlated value will be unity if the signal level has been normalised. However, when $m \neq 0$ we find that inside the summation we are multiplying pairs of values which aren't the same. Indeed, if the signal pattern is *efficient* we can't predict any one value in the series from any of the others. The offset therefore breaks the relationship that causes the $m = 0$ correlation to give a distinct positive value. The result is as if we were correlating to completely unrelated patterns!

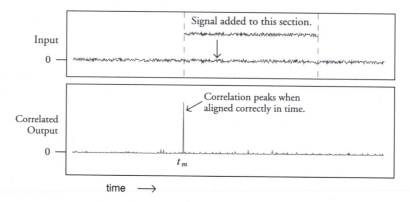

Figure 23.5   Sliding Correlation

The result is shown graphically in Figure 23.5 which plots $C\{N, p\}$ as a function of $p\Delta t$ below the original input. The 'hidden' signal only correlates constructively with the *Key* message pattern we are looking for when we arrange that we take the set of values starting where $t_m = p\Delta t$. Conceptually, we are essentially 'sliding the key pattern along the input looking for a match'. Hence this type of process is sometimes called *Sliding Correlation*. The clear peak when we start with the value for which $t_m = p\Delta t$ shows both that the sought pattern has arrived and also indicates <u>when</u> it arrived. (For clarity, figure 23.5 shows the square of $C\{N, p\}$ to indicate the SNR of the result.)

In practice, the above process may be quite tedious to perform as it involves repeated summations over products of large sets of values, which must also be redone for each of the possible message patterns in which we have an interest. Hence the whole process may become computationally intensive and take longer than we would wish. Fortunately, there are some more efficient ways to perform the same process. The most common of these is the use of an FFT-based method. This is based upon pre-computing the complex conjugate of the Fourier Transform of each library pattern and using these instead of the initial patterns. We then collect chunks of N samples from the input signal as they arrive, Fourier Transform them, multiply these values by our new library of transformed patterns, then inverse transform the result.

If one of the patterns we are looking for is present in the input, the result shown the same kind of peak we see in Figure 23.5. Although the details are more complicated in practice than described here, the system has one great practical advantage. It simultaneously tests for possible signals starting at N time offsets. This method is therefore often preferred as being faster and easier than the 'brute force' method of repeatedly just calculating each Correlation in turn, for every possible starting time.

In fact, there are a variety of ways we can search for signal patterns that are, in information theory terms, equivalent to the methods described in this chapter. For example, when building analog circuits, another common method has been based upon what is called a *Matched Filter*. This uses a filter which has been designed to give the maximum possible response when it is fed one of the patterns we are looking for. Parallel arrays or *Banks* of such filters can then be employed to quickly scan an input looking for signals. This approach is used less often these days as it has been overtaken by digital computations, but a numerical equivalent is still employed. In principle, however, both the FFT-based and Matched Filter based methods are equivalent to simple Sliding Correlation. The choice of method is simply for reasons of convenience, not for any abstract theoretical reason.

Summary

You should now understand how *Correlation* can be used to identify when a specific signal pattern has arrived, and can determine the time of arrival. It should also be clear how choosing an *Orthogonal Set* of patterns maximises our ability to decide which of them has arrived when the reliability of recognition is affected by noise. You should also now know what is meant by *Process Gain*, and that this increases with the number of samples in (or the duration of) a signal pattern. It should be clear that, as a result, Correlation provides a signal to noise enhancement which increases with the signal length.

Chapter 24

Time and frequency

24.1 The meaning of frequency

Many of the chapters in this book consider how we can accurately measure the shape or amplitude of signal patterns. In practice, we often also need to perform time or frequency measurements. These measurements can be performed in a variety of ways. However, before looking at some examples it is worth asking, just what do we <u>mean</u> by the 'frequency' of an input signal? The reason for this question is that, surprisingly, there is more than one definition of the term 'frequency' and we may get different results from a measurement depending upon which one the chosen technique assumes. There are three related problems which arise when we want to define and measure the frequency of a waveform. To understand these we can start by considering figure 24.1. This shows three signal pattern observations, each made over a finite duration, T_0.

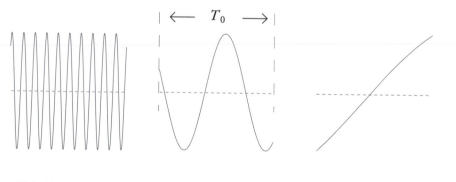

24.1a Many cycles in observed signal duration

24.1b Around one cycle in observed signal duration

24.1c Is this a part of a sinusoid at all???...

Figure 24.1   Three signal patterns of finite duration that may be sections from a sinusoid.

The waveform shown in 24.1a exhibits very clear signs of being a sinusoid with a fairly obvious frequency and amplitude. The presence of a number of repetitive cycles provides a strong argument for this. In comparison,

234

24.1b is less obviously a section of a sinusoid, although it still seems a reasonable deduction. Finally, with the signal pattern shown in 24.1c it isn't at all obvious that the shape is a section taken from a sinusoid since – if it is – we have so limited a section as to make the shape's nature unclear.

In practice, all observations are limited to a finite length or duration. However we tend to associate the concept of 'frequency' with sinusoids which, by definition, extend over an infinite duration. In Chapter 7 we saw how *Fourier Analysis* can be employed to obtain the *Frequency Spectrum* of an input waveform or signal pattern. This technique was applied using the assumption that we could always treat a pattern of duration, T, as being *Periodic* and that it would repeat itself after this time interval. Taking 24.1b as an example, this implies that the signal – if observed for much longer than T_0 – would look like the non-sinusoidal pattern shown in Figure 24.2.

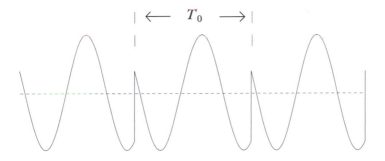

Figure 24.2 Extended waveform implied by assuming that the finite section shown in 24.1b repeats every T_0.

In the absence of any other information this result is fine since we would then have no reason to expect the 'chunks' shown in Figure 24.1 to actually be part of a sinusoidal wave at other times. Hence we would not need to worry that the three examples shown would all show quite different, and complicated, spectra when analysed using Fourier Analysis as applied in Chapter 7. However in each case the result has an 'enforced' periodicity imposed by the assumptions which then influences the results we obtain. This may conflict with any periodic behaviour inherent in the signal itself and lead to a misleading result in some circumstances. To avoid this problem, we can choose to make some alternative assumptions about the nature of the waveforms.

By looking at the patterns shown in figure 24.1 we can decide to assume that they are, indeed, all sections of a sinusoid. In some cases, instead of subjective recognition as the basis for such an assumption, we may have some extra information about the way the waveform was generated which tells us that it is reasonable to make this new assumption. Starting from this new basis, we can now hope to find a way of determining the relevant sinewave's frequency without being confused by the effects of the chosen finite duration, T_0, upon the observations. Note that when we make this alternative assumption we aren't actually varying the information content of the signals, just interpreting the content in a more appropriate way. This underlines that the assumptions we make, and the methods used for analysis, can alter the values we obtain as the results of a measurement. In effect, we have changed our minds about what quantity we are measuring.

The second problem which arises when we make frequency measurements is that the waveform may obviously be repetitive, and have a clear period, but <u>not</u> be sinusoidal. Examples of some waveforms like this are shown in figure 24.3.

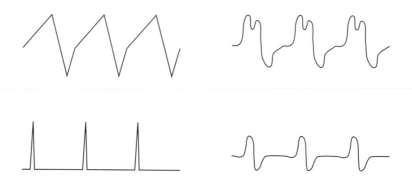

Figure 24.3 Various waveforms with the same repeat period.

As in the previous case, simply applying Fourier Analysis to obtain a spectrum on the assumption that the pattern is periodic over the entire observed interval may not provide a satisfactory measurement of the pattern's 'frequency'. In many cases like this we find it convenient to define the frequency of the waveform to be the number of times its shape repeats per second. i.e. we now define the frequency to be

$$f \equiv \frac{1}{t_p}$$

... (24.1)

where t_p is the repeat period of the waveform. This definition of a signal pattern's frequency is more general than our previous one which assumed sinusoidal behaviour. However there are still cases where we want to measure the frequency of a sinusoid, and will use the sinewave-based definition.

The third problem that affects frequency measurements is that the actual signal may not be perfectly periodic. Either as a result of drift, or random noise, or deliberate modulation, the frequency may change with time. The rest of this chapter will ignore deliberate frequency changes and only consider random noise effects as these can be expected to limit the accuracy of a frequency measurement.

24.2 Time and counting

Having established the way our assumptions can effect how a frequency value is determined, we can now examine some examples of various measurement techniques, starting with those which depend explicitly upon timing and waveform periodicity. *Counting* methods depend upon identifying some specific feature of the incoming periodic waveform and using this to *Trigger* a counting process. The chosen feature should only appear once per cycle. Hence by counting the number of times this feature *Event* occurs in a given time we can determine a frequency for the waveform. Using this method, we are defining the frequency to be the number of 'events (or cycles) per unit time' we observed.

An example of this method is illustrated in figure 24.4. The input waveform is passed through a *Comparator*. This has the task of comparing the input with a chosen *Decision Level* (sometimes called a *Cut* level). The output from the comparitor is binary – i.e. it provides one or the other of two output levels which just indicate whether the input signal is above or below the chosen decision level. This has the effect of converting the input waveform into a series of pulses. Usually, the comparitor is designed to output pulses whose voltages are compatible with digital logic gates (usually either TTL or CMOS).

Before proceeding to analyse the performance of this method it is worth noticing that we need to take care and ensure that we know how many times per cycle the input waveshape's level crosses the chosen decision level value, V_c, since this determines the actual number of events per cycle. Usually we try and ensure there is only one event in each cycle.

237

Multiple events per cycle do not matter provided we know how often they are occurring. However if we assume the incorrect number, the resulting frequency measurement will also be incorrect.

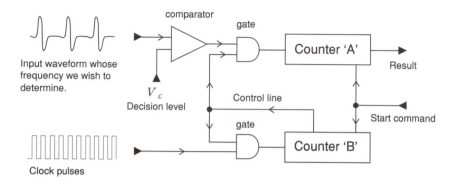

Figure 24.4   Frequency measurement based upon counting.

The system uses a pair of digital counters. The input to each of these is controlled by a *Gate*. Depending upon a control level supplied to these gates, we can either allow signals to reach the counters or block the inputs. When the gates permit, the lower counter, 'B', will count incoming pulses being supplied from a *Clock*. However, once this counter has received a set number of clock pulses, M, it changes the level of an output line, labelled as the 'control line'. This alters the action of the gates, blocking the entry of any more signal and clock pulses into the system. Counter 'A' will count the number of signal events (and hence the number of input signal cycles) allowed through by its gate.

The overall sequence for a measurement is therefore as follows:

- A 'start command is received by both counters. This clears the values they hold to zero.
- A count now starts as clearing counter 'B' caused its control line output to change, allowing both gates to pass the input signals.
- The count continues until counter 'B' has received M clock pulses. Once this occurs, the control line level changes and counting halts.
- Counter 'A' now stores the number, N, of signal events (cycles) it observed during the duration of the counting process.

Provided that we know the pulse-rate, f_{cl}, provided by the clock, and that this rate is stable, we can now read counter 'A' and say that N signal cycles occured in a measurement time interval of M / f_{cl}. Hence the measured frequency value for the input signal waveform will be

$$f = \frac{N}{M} \times f_{cl} \qquad \ldots (24.2)$$

Having established the basic method we can now assess the precision of any measured values we obtain using this method. This will clearly be affected, as usual, by the presence of noise. However before considering the effect of noise we should note some limitations inherent in the method. These arise from the quantised nature of the counting process.

The system counts whole events/cycles. Consider a situation where there happened to be, say, 1357·56 signal cycles during the observation interval. The counting would round this down to 1357. Hence the measured frequency would be in error due to the loss of the fractional part of the true value. Also, if the frequency changed before a remeasurement to, say, 1357·55, we would not notice as this value would also be rounded down. The inherent frequency accuracy of this measurement process is therefore such that we must expect a typical frequency accuracy of no better than

$$\delta f \approx \pm \frac{f}{N} \qquad \ldots (24.3)$$

This will be the case even if we could ensure a clock that was perfectly stable, running at a precisely known rate, and remove all random noise from the system. In practice we can seek to improve this measurement in two ways. The first, and most obvious, is to choose a larger value for M or a lower clock frequency, hence increasing the counted value for N. This increases the required measurement interval and we observe the input signal for longer than before. A similar improvement can be obtained by adding up and averaging a series of measurements as this also increases the total observation time.

A second, less obvious, potential improvement is to consider swapping over the signal and clock inputs and choose to use a very high clock frequency. This modified system counts how many <u>clock</u> cycles occur during a set number of input <u>signal</u> cycles – i.e. the inverse of the original process. The inherent accuracy is now determined to ± one clock cycle, not one signal cycle. Hence in this alternative arrangement we would obtain a measurement with a typical inherent accuracy of

$$\delta f \approx \pm \frac{f}{M} \qquad \ldots (24.4)$$

In practical cases we should use whichever frequency is the lowest to determine the counting interval, and maximise the number of counts of the other, in order to obtain the most accurate results.

24.3 Effect of noise on counting methods

For the sake of the explanations in this section we can assume that the counting method extends over many cycles so we can therefore neglect the rounding accuracy limit explained in the last section. Here we can consider the effects of noise upon frequency measurements obtained via counting or timing processes. A detailed analysis of this is complex so here we will adopt a simplified explanation which, nevertheless, gives the correct result in most cases of interest.

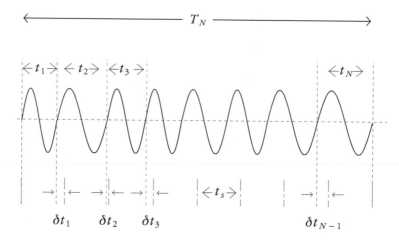

Figure 24.5 Observing the periods of a series of cycles.

Figure 24.5 is a schematic representation of the process of performing a timing measurement over a number of cycles. For the sake of clarity and simplicity we will only consider quasi-sinusoidal waveforms here. We then also use the zero crossings in the positive-going direction as our events indicating the start of each cycle period.

Before proceeding, it is worth bearing in mind that any periodic waveform can be regarded as having a spectrum that consists of a nominal *Fundamental* frequency, plus harmonic components. We could therefore choose to filter a waveform to remove the harmonics, leaving just the fundamental sinusoid which could then be processed as considered here.

This approach can, in fact, be useful in practice as it removes the noise power present at frequencies well away from the fundamental period/ frequency, hence potentially improving the SNR.

If the observed waveform were perfectly periodic, and no noise were present, we could expect to see a given number of cycles, N, occurring in an appropriate total observed interval, T_N. The waveform's period would then clearly be equal to $t_s \equiv T_N / N$. In practice, we can expect noise to randomly alter the instants when the zero crossings are observed. Hence instead of the observed period of every cycle being t_s it will have some value t_i which varies unpredictably from one cycle to the next. Random fluctuations in the time intervals between successive events is often called *Jitter*. These timing variations can be very important in some situation. For example, they may lead to problems in 1-bit DAC/ADC systems of the kinds considered in Chapter 22.

Whatever the source of the noise we can represent it as a series of timing errors for the observed events, δt_i. We can also represent the typical timing error level in terms of an amount Δ. Now when we just observe one cycle period we can expect each 'end' to have its apparent time altered by an unpredictable amount similar in magnitude to Δ. As a result since there are unrelated errors at the start and end of each cycle we would obtain an error in measuring the period of just one cycle that will typically be around $\pm\sqrt{2}\,\Delta$. i.e. The observed period would therefore typically be

$$t' = t_s \pm \sqrt{2}\,\Delta \qquad \qquad \text{...(24.5)}$$

where t_s represents the actual underlying signal period. Provided that the error is reasonably small compared to the period this leads to an uncertainty in the measured frequency taken using one cycle of around

$$\delta f \approx \pm f^2 \sqrt{2}\,\Delta \qquad \qquad \text{...(24.6)}$$

where f represents the nominal signal frequency. This frequency error is set by the relative levels of $\sqrt{2}\,\Delta$ and t_s, which is the observed time in this case. The above is therefore just equivalent to saying that

$$\frac{\delta f}{f} \approx \frac{\sqrt{2}\,\Delta}{t_s} \qquad \qquad \text{...(24.7)}$$

Consider now the effect of measuring the time taken for N cycles. We now ignore the timing errors of intermediate cycle locations and just see the errors in timing at the start and finish of this prolonged observation. So, since these timing errors are statistically similar to before, the level of frequency error changes to

$$\frac{\delta f}{f} \approx \frac{\sqrt{2}\,\Delta}{T_N} = \frac{\sqrt{2}\,\Delta}{N t_s} \qquad\qquad \text{...(24.8)}$$

This result tell us that making a measurement over N consecutive cycles gives an N-fold improvement in the accuracy of the frequency measurement. Where the timing errors are due to noise superimposed on a genuinely periodic waveform this is the correct result. However it isn't the right answer for situations where the jitter arises due to random fluctuations in the signal source which affect its behaviour.

The above analysis assumes that the actual signal generated by the signal source is perfectly regular, and each cycle has a period identical to all the others. This perfect regularity is just 'masked' by noise which is superimposed in between the creation of the signal and the observation process used to make a measurement. As a result, each observed zero crossing tends to remain within a typical $\pm\Delta$ from the actual instant the perfect regular underlying signal crosses the zero level. A similar result would apply for any other chosen event used to identify the start of each period cycle.

However when the signal source itself is affected by noise the period of its output may change randomly from cycle to cycle. Unless the source has some way to detect this and correct for the effect, the resulting time errors tend to accumulate incoherently. The result is that after N cycles the N-th zero crossing will not typically have a timing error of $\pm\Delta$, but $\pm\sqrt{N}\,\Delta$ due to the 'random walk' addition of all the earlier errors. In such cases the likely error leads to an uncertainly

$$\frac{\delta f}{f} \approx \frac{\sqrt{N}\,\Delta}{T_N} = \frac{\Delta}{\sqrt{N}\,t_s} \qquad\qquad \text{...(24.9)}$$

i.e. the probable accuracy of the frequency measurement only increases with $\sqrt{N}$, not N as would be the case where the noise is superimposed upon a perfectly regular signal pattern. Since this result implies a lower accuracy it is often wise to assume that it is so in order to avoid thinking that a measured result is more accurate than is really the case!

To obtain the above results we just have to count intermediate events (zero crossings) to ensure that we know how many cycles have occurred during the observed period, T_N. The precise instants when the intermediate events occurred were not noted. Some textbooks argue that we can obtain a more accurate result by recording each of the observed cycle periods, t_1, t_2, ..., t_i ..., t_N, and then taking their average. The argument presented is that all these values have independent errors which

242

can be reduced by the averaging process. Since this argument is put forwards in some texts it is perhaps worthwhile pointing out that it is incorrect. We can see why this is the case by looking again at figure 24.5 and imagining we had collected just a few cycle period lengths.

Take as an example, the error δt_2 which affects both t_2 and t_3. By looking at the figure we can make two important points. Firstly that the effects of δt_2 upon t_2 and t_3 have the same magnitude. Secondly, that they have opposite signs; i.e. if δt_2 makes t_1 longer by delaying the zero crossing, it shortens t_3 by precisely the same amount. If we choose to collect all the individual observed periods, t_1, t_2, t_3, etc, we calculate their average by performing two steps – adding up all the values, then dividing by how many values we collected.

The first step of the averaging process means that we obtain a total time

$$T_N = t_1 + t_2 + ... + t_N \qquad \qquad ...(24.10)$$

which is obviously identical to the time we would obtain by simply determining how long N cycles will take. The effects of all the intermediate error values vanish from this result because when we sum over all the individual times the effects upon adjacent values cancel out. As a consequence, in terms of obtaining an accurate result, there is no need whatsoever to record all the individual time periods since the averaging process gives us precisely the same answer as measuring the total time taken for N cycles.

The above does not mean that collecting the individual values is pointless. If we do not already know the value of Δ, collecting these values can be very useful in allowing us to assess the magnitude of the possible uncertainty in our measurement. By collecting a series of individual t_i values and calculating their spread we can estimate Δ and hence estimate the accuracy of measurement. However this process does not, in itself, alter the measured value we obtain.

24.4 Relationship between SNR and jitter level

In many cases the jitter arises due to the presence of a given superimposed noise. As usual with measurement processes the accuracy of measurement will depend upon the input SNR and the time taken for the measurement. The previous section is all in terms of the typical jitter level, Δ, so we should now establish how this is related to the input SNR. To understand their relationship we can use the example illustrated in figure 24.6. For the sake of our example this combines a sinusoid of amplitude, a, with some random noise, of rms level v_n. As usual, the effect of the noise will be to change the signal level in a manner that varies unpredictably from moment to moment.

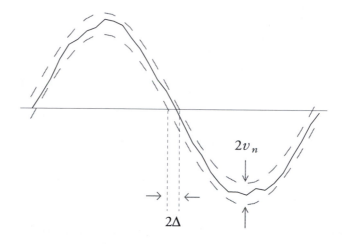

Figure 24.6 Jitter level caused by superimposed noise.

Since we are concerned with the typical or most probable effect of the noise we can imagine the underlying sinusoidal signal pattern as being 'surrounded' by a band of noise which typically extends $\pm v_n$ above and below the actual sinusoid. If we observe many cycles (e.g. by overlaying many cycles on an oscilloscope) this effectively 'blurs out' the sinusoidal pattern by this amount. Although this noise primarily alters the signal level, since the waveform crosses zero at a finite rate of change it also affects the instant of zero crossing.

For a sinewave

$$S = a \, \text{Sin} \{2\pi f t\} \qquad \qquad \dots (24.11)$$

the rate of change as the sinusoid crosses zero will be

$$\frac{dS}{dt} = 2\pi f a \qquad \qquad \dots (24.12)$$

A small change in level of $\pm v_n$ therefore alters the instant of zero crossing by an amount $\pm\Delta$ where

244

$$\Delta = \left[\frac{dS}{dt}\right]^{-1} \times v_n \qquad \text{... (24.13)}$$

i.e. we can say that the probable jitter level will be around

$$\Delta \approx \frac{v_n}{2\pi f a} \qquad \text{... (24.14)}$$

This result is an approximation which assumes that $a \gg v_n$ as it is based on assuming that the slope of the sinusoid remains essentially constant over the time interval Δ. In effect, we treat the zero-crossing waveform as locally being a straight line and then deduce the relationship between the values of Δ and v_n by assuming that their ratio tells us the slope of the line. Hence expression 24.14 is only reliable when the SNR is reasonably high. At lower SNR's, however, the counting method tends to fail to function reliably for the purposes of frequency counting. This is due to the noise tending to unpredictably alter the number of trigger events per cycle when the noise level is high. To obtain reliable counts we therefore require a reasonably high SNR. The above estimate can hence be regarded as being valid in most cases where the results of a count are likely to be accurate.

Expression 24.14 is in terms of amplitude levels. In general, the accuracy of measurements should be related to the SNR in energy or power terms as the result is then more general in its applicability. For a sinusoid the power level is proportional to $a^2/2$. We can therefore say that

$$\text{SNR} \equiv \frac{a^2}{2v_n^2} \qquad \text{... (24.15)}$$

Combining expressions 24.14 and 24.15 we obtain the result

$$\Delta \approx \frac{1}{2\pi f} \times \frac{1}{\sqrt{2 \times \text{SNR}}} \qquad \text{... (24.16)}$$

where SNR represents the input signal to noise ratio. By combining this result with expression 24.8 we can say that the resulting uncertainty of a frequency measurement will be

$$\delta f \approx \frac{1}{2\pi T_N \sqrt{\text{SNR}}} \qquad \text{... (24.17)}$$

This result is just an approximation but it serves to act as a guide to the effect of superimposed random noise upon the probable accuracy of frequency measurements based upon timing the period length of a number of cycles of a repetitive waveform. Note that, as we would expect from the basic concepts of information theory, the accuracy depends upon the observation duration, T_N, and the input signal to noise ratio.

245

In practice, frequency measurement methods based upon counting are actually comparisons since they determine the ratio of the number of cycles of the input signal to the number of cycles of the chosen clock which occur during the the observation time. This isn't a surprise as we established in the early chapters of this book that measurements are usually comparisons between a reference standard (in this case the time taken for a number of clock cycles) and the item we wish to measure. Once we are aware of the role of the clock in this process it becomes clear that any jitter or uncertainty in the period of the chosen clock will also tend to introduce some level of error or uncertainty into the result of the measurement process. When dealing with the methods considered in this chapter this has two consequences we have to bear in mind.

  • Any noise superimposed on the process of counting clock cycles will cause a jitter. The likely effect of this can be estimated using an approach similar to that used to deduce expression 24.17
  • Any error we make in defining or measuring the clock frequency will alter the results we obtain.

In many situations we can ensure that the effects of clock errors are relatively small and hence our measurements are limited by the signal SNR and the available obervation time, T_N. However we need to be aware of the need for a stable, well defined, clock to ensure this is the case.

Summary

You should now understand how *Counting* methods can be used to determine the period of a repetitive waveform. It should also be clear that the accuracy of a measurement will depend upon various factors which include the chosen clock rate, the signal's SNR, and the total time devoted to the measurement. You should also be aware that this kind of measurement is, like most others, a comparison, hence the stability of the clock, and the reliability with which we know its frequency will be important in ensuring an accurate result.

Chapter 25

Frequency measurement systems

25.1 Phase lock methods

Chapter 24 examined measurement techniques based upon counting methods. These are particularly useful where we can employ digital circuitry, and where we we want to measure the duration or period of a repetitive signal pattern. However they are not the best choice for every purpose. This chapter examines some of the other methods that are widely used to perform frequency and spectrum measurements. We will start by looking at the use of the *Phase Lock Loop* (PLL).

The *Counting* approach described in Chapter 24 relies upon using a suitable oscillator as a *Clock*. For accurate measurements to be possible this clock must have a well defined, and stable, frequency . In essence, the PLL approach turns the counting method on its head and seeks to adjust the clock frequency to *Synchronise* it with the incoming signal. If we then determine how much we have altered the clock frequency we can use this information as a measure of the frequency of the input signal. In practice, the term 'clock' is avoided when describing the oscillator that forms part of PLL systems since this word normally implies a stable frequency source. In most cases using electronics the alterations in the oscillator frequency are produced by applying a control voltage, hence the oscillator is referred to a *Voltage Controlled Oscillator* (VCO). Figure 25.1 is a schematic diagram showing the basic form of a simple PLL.

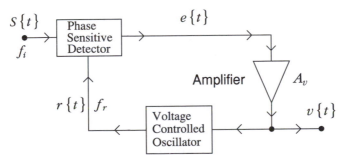

Figure 25.1   Simple Phase Lock Loop (PLL)

The system consists of a VCO, an amplifier of voltage gain, A_v, and a PSD connected together to form a closed loop. We have already examined the behaviour of *Phase Sensitive Detectors* in Chapter 16. The *Reference Frequency* for the PSD is supplied by the VCO. We can define the VCO's behaviour to be such that it produces an output whose frequency is given by the expression

$$f_r = f_0 + k_f v\{t\} \qquad \qquad ...(25.1)$$

where $v\{t\}$ is the voltage used to control the VCO. When $v\{t\} = 0$ the oscillator outputs a frequency, f_0, which is called it's *Free Running* frequency. i.e. f_0 represents the frequency the VCO will produce naturally when we make no attempt to alter its output.

From the basic properties of a PSD, the output, $e\{t\}$, can be defined as

$$e\{t\} \equiv k_p A \, Cos\{\theta\} \qquad \qquad ...(25.2)$$

where k_p is a measure of the gain of the PSD, A is the mean amplitude of the input signal, $S\{t\}$, and θ is the difference in phase at any time between the input signal and the reference, $r\{t\}$.

To understand how the system works we can start by considering the situation where the signal frequency happens to equal the VCO's free running frequency and the two happen to be in quadrature – i.e. we start by assuming that $f_s = f_r = f_0$ and the signal and reference differ in phase by 90 degrees. When the signal and reference are in quadrature, $\theta = 90°$, so $Cos\{\theta\} = 0$, and hence $e\{t\} = 0$. This means the VCO will continue to oscillate at f_0. As a result, unless we alter the input signal frequency this initial situation will continue.

Consider now what happens if the input signal's frequency changes slightly. At first the VCO output remains unchanged. The signal and reference frequencies are now different. Since frequency is the rate of change of phase, it follows that the phase difference between the signal and reference now starts to change, and θ departs from being 90 degrees. This alteration in relative phase means that the output from the PSD will also change and will no longer equal zero. The result is that, now $Cos\{\theta\} \neq 90°$, we get an amplified non-zero control voltage of

$$v_r\{t\} = k_p A A_v \, Cos\{\theta\} \qquad \qquad ...(25.3)$$

applied to the VCO, altering its output frequency to a new value

$$f_r = f_0 + k_f k_p A A_v \, Cos\{\theta\} \qquad \qquad ...(25.4)$$

The behaviour now depends upon arranging the loop and VCO to obtain the correct sign for the factors $k_f k_p A_v$. Provided that we ensure that this is such that the change in VCO output frequency has the same sign as the change in f_s we find that the VCO frequency now alters to 'follow' the change in input signal frequency. This change will continue whilst there is any difference between the two frequencies. This is because a continued difference in frequency means that the phase difference is changing, altering the PSD output that controls the VCO. However once the VCO frequency has changed enough and become equal, once again, to the signal's new frequency their relative phase difference becomes steady again. The result is that the system eventually settles at a new equilibrium where the signal and VCO frequencies are the same once more. However there is now a new difference in phase, maintained at whatever value is required to 'push' the VCO output to $f_r = f_s$ and preserve this equality.

The result is that the loop causes the VCO output frequency to *Track* the signal frequency. The system is, in fact, a *Feedback Loop* (hence the word 'loop' in its name). Most control loops feedback a voltage, but this one feeds back a frequency, and always tries to adjust this frequency to match that of the input. Provided that we know the values of k_f and f_0 we can now expect that

$$f_s = f_r = f_0 + k_f v\{t\} \qquad \qquad \dots (25.5)$$

Hence by observing the voltage, $v\{t\}$, we can determine the signal frequency, f_s.

PLL's are often used as *Frequency Demodulators* in communications systems since the output voltage will tend to vary in sympathy with any changes in f_s as time passes. PLL's are widely used in radio and microwave applications and lend themselves well to being incorporated into integrated circuits. Their main drawback is that we must have reliable knowledge of f_0 and k_f in order to convert the observed output voltage, $v\{t\}$, into a frequency value. We also have to take care in cases where the signal frequency fluctuates. When this happens, the output $v\{t\}$ will then, temporarily, become 'wrong' (i.e. incorrectly indicates the frequency) until the difference between f_s and f_r allows the phase difference between them to shift to a new level that alters $v\{t\}$ to bring them back to being equal again.

The system has a finite *Response Time* which determines upon how quickly it can react to changes in the input frequency. In some cases this can be useful as it means the output value will be averaged over a period of time,

249

smoothing away any brief frequency fluctuations. In fact, some counting systems exploit this behaviour and use a PLL to 'clean up' the input signal. In these systems the input signal is used to drive a PLL and a counter is used to count the VCO output, <u>not</u> the signal. Any swift temporary fluctuations in the input signal will tend to be smoothed away by the time constant chosen for the PSD. Averaged over a long enough time the VCO output has the same frequency as the signal, so can be counted in its place. The advantage is that the SNR presented to the counter has been improved by using the PLL to filter away swift variations. The disadvantage of the PLL is that we can't use it to detect or measure changes that are too swift for it to be able to respond.

25.2 Resonators and filters

Systems based upon Phase Lock Loops are now widely used in electronics as the circuits required work well and are easy to manufacture. However, it isn't always possible to employ a PLL. For example, it is currently impractical to manufacture a conventional electronic system which acts directly upon very high frequency signals – e.g. at the frequencies of visible light. Even at lower frequencies it is sometime more convenient to use other methods that were in widespread use before the development of the PLL. One of the most common alternatives is the use of some form of filter or device whose behaviour is inherently frequency sensitive, and then use that sensitivity as a route to performing frequency measurements. Figure 25.2 shows a couple of examples of the kinds of electronic circuit often used for this task.

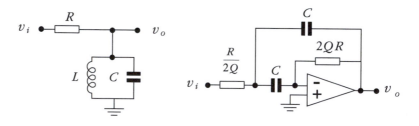

25.2a Passive bandpass filter   25.2b Active bandpass filter

Figure 25.2   Examples of bandpass filters

Figure 25.2a illustrates a passive *RCL Bandpass Filter*. Figure 25.2b shows an active filter system which performs in a similar manner. The practical advantage of 25.2b is that it does not require an inductor. This makes it smaller, cheaper, and gives better defined performance. However, the need for an amplifier means it requires power, and limits its use to frequencies where suitable amplifiers are available.

Although the systems employed at optical frequencies are physically constructed in quite differently to the above they are used in similar ways. Here we can concentrate on using electronic examples, but it should be remembered that equivalent arrangements can perform the same functions in other frequency ranges. For example, the *Fabry-Perot Resonator* is often used at mm-wave and optical frequencies as a filter in much the same way as the electronic arrangements considered here.

For the sake of example, lets look at the behaviour of the circuit shown in figure 25.2b. Using the standard methods of complex circuit analysis this can be shown to have a voltage gain

$$A_v\{f\} \equiv \frac{v_o}{v_i} = \frac{-j\,2Q^2 ff_0}{f_0^2 - f^2 + jff_0/Q} \qquad \text{... (25.6)}$$

at a (sinusoid) frequency, f, where

$$f_0 \equiv \frac{1}{2\pi RC} \qquad \text{... (25.7)}$$

represents the filter's *Resonant Frequency*, and Q is the quantity usually referred to as the filter's *Quality Factor*.

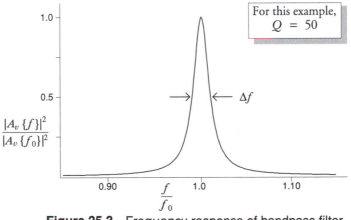

Figure 25.3 Frequency response of bandpass filter

251

Figure 25.3 shows a plot illustrating how the power gain of an example of this type of filter varies with frequency. The graph is normalised in terms of the circuit's resonant frequency and its power gain at that frequency. You can see that the circuit's gain has a peak value at f_0. The shape of the response shows that the gain is only greater or equal to half its peak value over a limited range of frequencies

$$\Delta f = \frac{f_0}{Q} \qquad \qquad \text{... (25.8)}$$

centred upon the resonant frequency.

There are two ways we can imagine this system being employed. These differ as a result of the assumptions we make about the nature of the incoming signal. The first case is where we have reason to assume that the signal essentially consists of a single frequency component, although the frequency may fluctuate slowly over a limited range. In this situation the system is usually referred to as a *Frequency Discriminator*. The second case is where we assume the signal has power spread out over a wide range of frequencies. Here it is the bandpass filtering property – i.e. passing power in a selected band of frequencies – which is used.

Let's start with the frequency discriminator. Here we try to arrange for the resonant frequency to differ from the nominal signal frequency, f_r, by an amount similar to Δf. This is illustrated in figure 25.4.

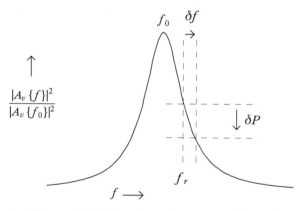

Figure 25.4 Bandpass frequency discriminator

Here we have arranged for f_r to be equal to $f_0 + \Delta f / 2$. Hence when at this frequency the output power emerging from the filter is half the value it would have if the signal were at the resonant frequency, f_0. A small

252

change in the signal frequency, δf, will cause a corresponding change, δP, in the output power level. We can therefore use the filter as a Frequency-to-Voltage convertor and deduce a frequency change when we observe a change in the output power level.

The *Conversion Gain* or *Sensitivity* of the filter can be defined as the ratio between the observed change in output power level and the corresponding change in frequency. The value of this ratio will depend upon the shape of the filter's frequency response curve and the location we choose for f_r relative to f_0. We can, however estimate the approximate sensitivity by noting that the power output falls by a factor of 2 as we change the frequency from f_0 to $f_0 + \Delta f / 2$. We can therefore say that, approximately, the magnitude of the observed change in power will be given by the expression

$$\delta P \approx P_0 \frac{\delta f}{\Delta f} \qquad \ldots (25.9)$$

i.e. the sensitivity will be approximately

$$\frac{\delta P}{\delta f} \approx \frac{P_0}{\Delta f} = Q \frac{P_0}{f_0} \qquad \ldots (25.10)$$

As usual, in practice our ability to detect a change in frequency will be limited by the smallest change in the power level we can observe. Hence to obtain a high sensitivity we would wish to arrange for the product $Q P_0$ to be as large as can be arranged. However we must take care not to employ too narrow a filter – i.e. have an excessively large Q value. This is because the above result will only be approximately correct for small changes – i.e. when δf is somewhat smaller than Δf. Increasing Q would increase the sensitivity, but also reduces the frequency range over which the bandpass discriminator will work as expected. For example, in the case illustrated the signal frequency is greater than f_0. A small reduction in signal frequency will tend to cause the output power to rise. However if the signal frequency falls below f_0 the output power reaches a peak value and then falls. Similarly, once the signal frequency is more than a few times Δf away from f_0 the output will essentially be zero whatever the frequency since the filter is rejecting power at these frequencies.

In addition to having a limited useful range, determined by the chosen Q value, the system has two significant drawbacks. Firstly, the actual relationship between the output power and the signal frequency is non-linear. Hence if we assume a linear relationship (as implied by expression 25.10) we will obtain an incorrect or distorted output measure of the true frequency. To avoid this we must either use a more precise knowledge of the filter's properties to convert power observations into frequency measurements, or restrict any frequency changes to be much less than Δf. By only using a very small portion of the filter's power/frequency curve we can approach linear behaviour as any short section of a small curve approaches a linear tangent provide we use a small enough section.

The second problem is indicated by the fact that P_0 appears in the expression for the sensitivity. This warns us that the output power level depends upon the actual input signal power level. Unless we take care we will therefore find that any unexpected changes in the signal power will be interpreted, incorrectly, as being due to frequency changes. The simplest way to deal with this is to monitor the input signal power and then compare this with the power level emerging from the filter. This can also be done in optical systems by simultaneously measuring both the transmitted and reflected power levels from the filtering element and summing these to deduce the input power level. Some electronic systems also use two or more filters, tuned differently, and compare their outputs to give an indication of the input power and also help correct some of the distortions produced by the non-linear curves of the filters.

The second method for using the bandpass filter is simpler to explain as it depends upon the filter's property of passing signals in a range of width Δf centred upon f_0 whilst rejecting signals at other frequencies. We can therefore just measure the output power and deduce that this represents a measure of the power of the input signal's frequency components in this range. By employing a set of filters, each tuned to a different band, and measuring the output powers they provide we can build up a spectrum of the input signal. This method is useful when dealing with wideband signals rather than simple periodic waveforms. The results may sometimes have to be interpreted with care as the above assumes that each filter has a 'top hat' shape – i.e. has a uniform gain over a range, Δf, and perfect rejection of frequencies outside this band. In reality the filter shape means some of the 'in band' power is lost, and some power at other frequencies may 'leak' through. When dealing with smooth spectral distributions this isn't a major concern, but it may have an effect in other circumstances.

A series of such filters, tuned to adjacent frequency bands, and employed in this way is called a *Filter Bank*. Such a system allows us to monitor the power levels in a series of bands simultaneously. A common alternative is to 'sweep' the tuning of a single filter so as to vary the band where power is passed through, and then note how the observed power varies as the filter is swept. This is the method used for most RF *Spectrum Analysers*. The disadvantages of this method are that it may miss changes that occur whilst sweeping is taking place, and that it only observes power at each frequency for a small fraction of the time. Hence it is inefficient in SNR terms as signal energy can go unobserved, and there is also a risk that power fluctuations may be appear, falsely, as spectral features.

25.3 Fourier transform spectroscopy

We have already encountered Fourier Transformation in earlier chapters. This technique is widely used to make signal frequency and spectrum measurements. The classical optical instrument employed for this approach is the *Michelson Two-beam Interferometer*, although many other forms of two-beam interferometer are used. At microwave or mm-wave frequencies equivalent systems are sometimes used, but the signals may be carried along a variety of guiding structures (waveguides, stripline, etc).

Although these arrangements vary in their details we can explain how they all work in terms of a 'generic' form for a two-path interferometer as illustrated in Figure 25.5. This shows a system constructed using a pair of symmetric 50:50 power-splitter/combiner elements, linked via two paths. The relative lengths of these paths, Z_1, and Z_2, may be varied. The two outputs of the second splitter are directed onto a pair of power detectors.

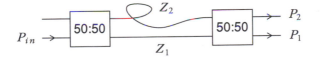

Figure 25.5 Schematic of 2-path interferometer.

The splitter/combiners can be assumed to be free of loss and balanced in the way they behave. This means that the sum of the fields (or voltages) exiting each equals the total field that arrives. It also means that the total input and output powers must be the same. At first glance this apparently leads to a paradox since when we have two exit voltage levels, $V_1\{t\}$, and $V_2\{t\}$, produced by splitting an input, $V_{in}\{t\}$ it means that we must simultaneously ensure that both of the expressions

$$V_1\{t\} + V_2\{t\} = V_{in}\{t\} \qquad \ldots (25.11)$$

$$[V_1\{t\}]^2 + [V_2\{t\}]^2 = [V_{in}\{t\}]^2 \qquad \ldots (25.12)$$

must be correct. This is a consequence of the power of a signal being

255

proportional to the square of its voltage (or to the field of a distributed field). These two requirements can only be satisfied together by allowing that the split fields have their relative phases altered. For an input component at some frequency, f, the outputs will have their relative phases so that they emerge with a difference in phase of 90 degrees relative to each other – i.e. ±45 degrees relative to the input.

The system shown in figure 25.5 has four *Ports* (ways in or out) by which signals may enter or leave. In principle we can use these as we wish, but here we can consider allowing signals to enter just via one port, and then observe the results that exit via the pair of ports at the other end of the system. Taking the above point about phases into account a detailed analysis of the system will establish that when we input a power, $P\{f\}$, at a frequency, f, the powers exiting at this frequency will be

$$P_1\{\Delta, f\} = \frac{P\{f\}}{2}\left[1 + \mathrm{Cos}\{2\pi f\Delta/c\}\right] \qquad \dots (25.13)$$

$$P_2\{\Delta, f\} = \frac{P\{f\}}{2}\left[1 - \mathrm{Cos}\{2\pi f\Delta/c\}\right] \qquad \dots (25.14)$$

where $\Delta \equiv Z_1 - Z_2$ represents the *Path Difference* value – i.e. the difference in the lengths of the two signal paths within the system – and c is the velocity of the signals propagating through the system. (In an optical system this would, of course, be the speed of light.)

Where the input signal consists of a set of components at various frequencies we can treat these as propagating independently through the system and hence the resulting output power is just the sum of what we would get according to expressions 25.13 & 25.14 for each spectral component. We can represent these overall output powers as $P_1\{\Delta\}$ and $P_2\{\Delta\}$ to remind us that the result depends upon the chosen value of Δ.

It is usual practice to call the manner in which the observed output powers vary with the path difference value the *Interferogram* of the input signal. To employ the system as efficiently as possible we would wish to use both of the detected output power levels as part of a measurement process. When trying to make frequency measurements we also usually find it useful to normalise any measurements against the input power level to try and prevent input power changes from accidentally being interpreted as being due to frequency changes. As a result it is convenient to mathematically define a normalised interferogram pattern as being

$$I\{\Delta\} \equiv \frac{P_1\{\Delta\} - P_2\{\Delta\}}{P_1\{\Delta\} + P_2\{\Delta\}} \qquad \dots (25.15)$$

The interferogram shape we obtain by observing how the output powers vary with the path difference now provides information about the spectrum of the input signal.

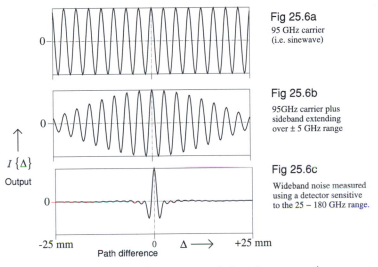

Fig 25.6a
95 GHz carrier
(i.e. sinewave)

Fig 25.6b
95GHz carrier plus
sideband extending
over ± 5 GHz range

Fig 25.6c
Wideband noise measured
using a detector sensitive
to the 25 – 180 GHz range.

$I\{\Delta\}$
Output

-25 mm   0 $\Delta \longrightarrow$ +25 mm
Path difference

Figure 25.6   Some example interferogram shapes.

Figure 25.6 shows three examples which we can use to illustrate the relationship between the shape of the interferogram and the spectrum of the input signal. Figure 25.6a represents the output we would obtain from an input source which is generating one single (sinusoidal) frequency component. In this example the source's frequency is 95 GHz. We can see that – as implied by expressions 25.13 to 25.15 – the interferogram shows a sinusoidal variation with Δ. The observed period of this sinusoidal variation of the interferogram shape will be λ, the wavelength of the input signal. Hence by collecting the interferogram, and determining its period, we can measure the signal's frequency provided that we know the relevant propagation velocity, c.

Figure 25.6b has a similar shape and shows a pattern that looks almost sinusoidal. However we can see that the amplitude of the apparent sinusoid tends to decline as we move away from zero path difference. In this case the input signal has a spectrum which contains power over a modest range of frequencies centred upon 95 GHz. To understand this pattern we can consider again the implications of expressions 25.13 and 25.14. By looking at these we can see that whatever the input signal

257

frequency, at zero path difference we would have $P_1 = P_{in}$ and $P_2 = 0$; i.e., since it is normalised, $I\{0\} \equiv 1$ <u>whatever</u> signal we choose to input.

As we move away from the zero path difference setting, each frequency component in the spectrum varies its contribution to the output with a periodicity that varies according to its wavelength. The total interferogram represents the sum of each of the sinusoidal contributions they make, each with its own wavelength. At zero path difference (ZPD) these contributions are all 'in step' and give us a peak total output for $I\{\Delta\}$. As we move away from ZPD these contributions shift out of step and tend to interfere with each other. When the spectral components in the signal all have similar frequencies their contributions to the interferogram all have similar wavelength, hence this movement out of step only manifests itself when we move a fair way away from ZPD. Hence we see an apparent sinusoid whose amplitude decays away from the ZPD setting.

The example shown in figure 25.6c can be understood by taking this argument further. Now the signal spectrum has components over quite a wide range of frequencies. The resulting interferogram still has a peak at ZPD, but it quickly falls away to almost nothing when we move away from the zero path difference. Here the width of the central spike of the shape is a guide to the observed bandwidth of the signal. The wider the range of frequencies detected, the narrower the spike will be as a result of the contributions rapidly getting out of step as we move away from the ZPD.

Before moving on to the considering how Fourier Transformation can be applied to obtain spectral information, it is worth noting two useful points that sometimes go unnoticed. The first is that the interferogram shows a pattern that only depends upon the <u>detected</u> components of the input signal. Once said, this should be obvious, but it has an interesting result. An interferogram of the general form shown in figure 25.6c can often arise as a result of making observations with an input signal that is actually wideband (e.g. thermal) noise. The spectrum we see in this case is a signature of the frequency response of the detectors and the actual interferometer system we used to make the observation. Interferometers work by exploiting *Coherence*. We don't normally regard random noise as being coherent. Here, however a level of coherence is *Imposed* by the finite bandwidth of the system used to make the observations. The example shown in figure 25.6c assumes wideband noise but using detectors that only responds to power over a specific range. In this case its is the characteristics of the detectors that determine the shape of the interferogram.

The second point to note is that the above assumes that we are only permitting signal power to enter via one of the two possible input ports shown at the left side of figure 25.5. In practice we may also have a quite different signal entering simultaneously via the second input. When this occurs the interferogram we observe will depend upon the <u>difference</u> between the power spectra of the two input signals. This behaviour is a result of the symmetry of the system. If we had chosen to use the alternative input for all the above explanations the results would have been the same except for swapping over P_1 and P_2 and thus inverting I.

In most practical cases we are likely to find that some signal power – e.g. thermal noise from a background – will enter the second input port even if we do not wish this to happen. We therefore need to consider this possibility when making measurements as it may affect the results. Alternatively, we can choose to deliberately make use of this symmetric property and employ the interferometer for *Nulling* measurements. Here we can deliberately inject a controlled signal into one input and seek to adjust this so that, when combined with the signal from the other input, the result is a 'flat line' interferogram where $I\{\Delta\}$ is zero at all path differences. Once this occurs we can deduce that the controlled and uncontrolled signals have identical power-frequency spectra within the bandwidth the system can observe. Hence if we know the details of one, we have determined the details of the other without having to know many of the details of the actual measurement system.

The process by which a two-path interferometer is normally employed to make spectral measurements can now be understood as a simple application of the Fourier Transform (FT) methods described in earlier chapters. The usual process starts by sweeping the path difference over a suitable range of values and recording the interferogram pattern this produces. In theory we can choose any range we wish, but in practice the most sensible choice tends to be symmetric about the ZPD. The result is that we have now observed an input signal pattern, $I\{\Delta\}$, over a known interval, $-D \leqslant \Delta \leqslant D$, where D represents the range either side of ZPD that has been recorded.

In theory there is no need to make a symmetric measurement as the interferogram pattern should be such that $I\{\Delta\} = I\{-\Delta\}$. In practice, however, we may not have accurate pre-knowledge of the ZPD location and there may be system imperfections which distort the interferogram shape. We can generally detect and correct these by comparing the patterns either side of the ZPD and noting any unexpected asymmetry.

The input observation now forms a record of length $2D$ which we can expect to contain some superposition of sinusoidal contributions with wavelengths and amplitudes related to the frequencies and powers of the spectral components of the input signal. By Fourier Transforming the interferogram pattern we calculate the nominal spectrum. This process is subject to the same limitations and assumptions we considered in earlier chapters. The periodic assumption means that we obtain a spectrum in terms of a series of frequency components whose wavelengths are

$$\lambda_i = \frac{2D}{i} \qquad \qquad \text{... (25.16)}$$

where $i = 1, 2, 3, \ldots$ i.e. any positive integer up to some limiting value. This means we have a spectrum which essentially tells us the power levels in a series of bands centred upon the frequencies

$$f_i = \frac{ic}{2D} \qquad \qquad \text{... (25.17)}$$

For the interferogram (and hence the spectrum) to form a complete record of the input we must have enough samples to satisfy the sampling theorem. Hence if the highest detectable signal frequency is f_{max} we must ensure that more than

$$N_{max} = \frac{2Df_{max}}{c} \qquad \qquad \text{... (25.18)}$$

uniformly spaced samples are taken over the range $-D \leqslant \Delta \leqslant D$.

In practice the transformation is normally performed using an FFT method and it is therefore often convenient for the number of samples recorded to be a power of two. However this is not required for any theoretical reason, nor is it always vital to take samples uniformly spaced. These arrangements are purely for practical convenience. It is also worth bearing in mind that we do not have to perform a Fast Fourier Transform (FFT). We could use some analog arrangement (e.g. a set of electronic filters) to analyse the interferogram. Digital sampling and FT methods have become common simply because they are flexible, and convenient to use.

The conventional FT approach is based upon assuming that we can regard the signal spectrum as a set of components periodic in the observed interval ($2D$ in this case). In general this is fine as the resulting spectrum contains all the input signal's pattern in a useful form. However – as discussed in section 24.1 – where we have some other knowledge or expectation, we can analyse the interferogram equally well on an alternative basis.

260

A common example is where we have reason to feel sure that the signal source is producing a close approximation to a sinusoidal output. i.e. it has a spectrum which is confined to a frequency range narrow enough for us to regard it as being at a single frequency. We could then take the interferogram pattern and apply some curve fitting technique to find a best fit to a sinusoid. (If the interferogram is symmetric this would be a cosine in Δ.) Another alternative is that we could use the interferogram, plus a series of pulses from the system scanning the path difference as the inputs to a counting system of the general type described in Chapter 24. This forms what is sometimes called a *Fringe Counting* approach. Many interferometers employ some form of stepping system to change the path difference, or employ an encoder that gives pulses that are regularly spaced along the range of path difference. These can be used as the 'clock' for comparison with a periodic interferogram shape.

The above non-Fourier methods are only useful when we are confident that the signal is periodic. However they have two very useful properties when this is the case. Firstly, they are not subject to the *Resolution Limit* of a standard Fourier method. This stems from the way the FT only gives results at a series of specific frequencies, spaced apart by an interval,

$$\delta f = \frac{c}{2D} \qquad \qquad \ldots (25.19)$$

As a consequence, when we employ Fourier methods to compute a spectrum for a signal which may contain many components spread over a wide bandwidth, the result does not normally allow us to unambiguously distinguish spectral features with a resolution finer than δf. The sinusoid-fitting and fringe counting methods are employed in situations which are should not subject to this limitation as the signal should be periodic. Provided that the measurements are made with a high enough SNR, etc, we can then hope to determine the frequency of a periodic input more accurately than δf. The fringe counting method can also provide results relatively quickly because the counting process may proceed while the interferogram is recorded, whereas an FFT computation takes some time and normally cannot start until the complete interferogram record has been collected.

Despite the above, since Fourier Transformation does not lose any information we can post-process an FT computed spectrum to obtain higher frequency accuracy provided that we are confident that the input signal is periodic. In effect, this means we fit a shape to the spectrum and find its 'peak' even when the peak does not coincide with one of the frequency, f_i, values where the FT computed a nominal power level.

This method would ultimately give the same results as fringe counting or fitting a cosine to the interferogram, but is rather more 'around the houses' in computational terms. The critical point here is that we can only apply these methods when we are confident that we know that the input signal is periodic. When this is the case we can apply sinusoid fitting, or fringe counting, or FT and peak/shape fitting as we prefer and the results should be the same if the process was carried out with care.

When we have no pre-knowledge of the signal's spectrum we cannot reliably apply these fitting or counting methods as any results they produced would probably be meaningless. In general therefore, where we have no other information, we must normally accept that the resolution of measurements upon signals with complex spectra will probably be limited to the value of δf given by expression 25.19.

Summary

You should now understand how a simple *Phase Lock Loop* can be used to make frequency measurements and how it can be used to detect changes (or modulation) in the frequency of a periodic signal. You should now also know how a resonant filter can be used, either to select power in a chosen band for measurement, or as a frequency sensitive element to detect frequency changes. It should also be clear that a filter should use used with care as fluctuations in the input signal's power may appear as frequency changes, and that to avoid this we may need to monitor the signal power. Finally, you should now understand how a two-path (or two-beam) interferometer can be used to make spectral measurements. That this is usually done using Fourier Transform methods, but that other approaches can be useful when making measurements upon a signal we feel confident is periodic.

Appendix 1

Solutions to numerical questions

Chapter 2

Question 2. The total voltage range is 4 V (+2 to –2). The noise blurs out any reading over a range of 2 mV (+1 mV to –1 mV). Hence we can divide the total range up into 4 V/2 mV = 2000 distinct bands. Since each band requires its own symbol this means we need 2000 symbols to cover or describe all the distinguishable levels. An n-bit ADC produces an n–bit binary word for each sample. There are therefore only 2^n possible values it can indicate. When $n = 11$, $2^n = 2048$.

Question 3. An oscillation at some frequency, f, means the level moves back and forth between a given maximum and minimum level during each cycle, lasting a time $1/f$. Since the level has to both rise and fall it follows that it must switch from one level to the other in half this cycle time. Hence a maximum frequency capability of 150 kHz means than the channel's response time must be $1/2f = 3.3\ \mu S$.

Question 4. We have to take a new sample as soon as a response time has elapsed since the last one. (Taking samples more often is a waste of effort since the level hasn't had time to change significantly. If we take them less often we risk missing something.) Hence we have to take $1/3.3\ \mu S$ samples per second — i.e. 300,000 samples/second. This means we'll collect 3 million samples during a 10 second message. Each sample can contain 11 bits worth of information. So we can get 33 million bits of information from the message.

Chapter 3

Question 3. Using equation 3.11 we can say that the thermal noise from a 10 kΩ resistor is such that

$$\langle e_n^2 \rangle = 4 \times 1.38 \times 10^{-23} \times 300 \times 1 \times 10000$$

$$= 1.65 \times 10^{-16} \qquad \text{V}^2/\text{Hz}$$

We can now use equation 3.14 to work out the noise per unit bandwidth entering the amplifier. Choosing $R = 10\ \text{k}\Omega$ and $R_{in} = 22\ \text{k}\Omega$ we can say

263

$$N = \frac{1.65 \times 10^{-16} \times 22000}{(10000 + 22000)^2} = 3 \cdot 54 \times 10^{-21} \qquad \text{W/Hz}$$

Question 4. The maximum power transfer will occur when the source resistance and input resistance are equal in value, so we require $R_{in} = R = 10 \text{ k}\Omega$. We can either use the same equation as before, or use 3.15 to say that the noise power spectral density will then be

$$N = \frac{1.65 \times 10^{-16}}{4 \times 10000} = 4 \cdot 12 \times 10^{-21} \qquad \text{W/Hz}$$

Chapter 4

Question 1. The chance of a '1' being correctly transmitted is C_1 and the chance of a '0' being correctly transmitted is C_0. The question tells us that we start with the values $V_1 = 4 \cdot 5$, $V_0 = 0 \cdot 5$, and $\sigma = 1 \cdot 5$. From the question we can set the decision level to be $V' = 2 \cdot 5$. (All values in volts.) Using expressions 4.7 and 4.8 we can therefore say that

$$C_1 = \frac{1}{2}[1 + \text{Erf}\{1 \cdot 88\}] \qquad ; \qquad C_0 = \frac{1}{2}[1 + \text{Erf}\{1 \cdot 88\}]$$

Using the expression for Erf given in the question we obtain the result

$$C_1 = C_0 = 0 \cdot 996$$

There are $N_1 = 2000$ '1's and $N_0 = 2000$ '0's, hence the total number of bits correctly received will be

$$N_{ok} = N_1 C_1 + N_0 C_0 = 3984 \text{ bits}$$

Question 2. Changing the decision level to $V' = 3\text{V}$ means that

$$C_1 = \frac{1}{2}[1 + \text{Erf}\{1 \cdot 41\}] \qquad and \qquad C_0 = \frac{1}{2}[1 + \text{Erf}\{2 \cdot 36\}]$$

i.e. $N_{ok} = N_1 C_1 + N_0 C_0 = 2000 \times 0 \cdot 9769 + 2000 \times 0 \cdot 9996 = 3953$ bits. Changing to $V' = 1 \text{ V}$ means that $C_1 = 0 \cdot 9999$ and $C_0 = 0 \cdot 7470$ so the correctly received bits will probably then be $N_{ok} = 3494$. (Note that in each case this is only the 'most likely' answer since the actual number of errors depends upon the actual noise pattern during transmission.)

Chapter 5

Question 1. 1024 symbols $\times$ 2 bits/symbol = 2048 bits of information.

Question 2. Equation 5.13 tells us the amount of information in a typical message. In this case $n = 512$, $P_1 = P_2 = P_3 = P_4 = 0.125$, and $P_5 = P_6 = 0.25$, so for a typical message

$$H_{typical} = \sum_{i=1}^{M} -NP_i \log_2\{P_i\}$$

$$= -512 \times \left(4 \times 0.125 \times \log_2\{0.125\} + 2 \times 0.25 \times \log_2\{0.25\}\right)$$

If your calculator doesn't have a $\log_2$ button you can make use of the relationship

$$\log_2\{x\} = 3.33 \times \log_{10}\{x\}$$

to work out this means that $H_{typical} = 1280$ bits for a typical 512 symbol message.

Equation 5.16 tells us

$$H_{specific} = \sum_{i=1}^{M} -A_i \log_2\{P_i\}$$

is the amount of information in a <u>specific</u> message which contains A_i occurrences of the each symbol, X_i. For the case described in the question this means that

$$H_{specific} = -300 \times \log_2\{0.125\} - 100 \times \log_2\{0.125\} - 112 \times \log_2\{0.25\}$$

i.e. $H_{specific} = 1424$ bits for this particular message.

Chapter 6

Question 1. In this case $E = 0.1$ and $C = 0.9$ for each bit.
The message is $N = 10,000$ bits long. For 'tell me once' $C \times 10000 = 0.9 \times 10000 = 9000$ bits are likely to arrive without errors. i.e. $E \times 10000 = 1000$ of the received bits can be expected to be incorrect.

Using 'tell me three times' we can expect around $N C^3 = 7290$ of the bits to agree in all three copies because they are all error free. The number of message locations where two copies are correct and just one is erroneous will be $3N E C^2 = 2430$. These errors can be detected and corrected, and the recovered bits added to the 7290 which arrived without any errors to produce a correctly received (including corrected) total of 9720.

The number of times two bit locations are in error and just one is correct will be $3NE^2C$ = 270. Hence there will be around 270 occasions when we will see that an error has occurred, but will make the wrong correction decision. These errors have been detected, but not corrected properly. Only on NE^3 = 10 occasions will all three copies agree because they are all erroneous. As a result the final 'corrected' message is — following 'tell me three times' transmission — likely to contain around 270 + 10 = 280 undetected errors.

Question 2. Arrange the 16 bits into a 4 × 4 square. This contains all the initial information. Adding one parity bit per column and one per row produces an extra 4 + 4 = 8 bits. Hence the total number of bits, including parity is 16 + 8 = 24. The efficiency is defined as the ratio

$$\text{initial/total} = 16/24 = 0{\cdot}666$$

The redundancy is defined as 1 minus the efficiency, i.e. $0{\cdot}333$ in this case.

Chapter 7

Question 1. The bandwidth of the signal coming from the microphone is 18,000 - 10 = 17,990 Hz. We therefore need to take at least 2 × 17,990 = 35,980 samples/second to make a complete record. The song is 3 minutes (i.e. 180 seconds) long, so the total number of samples required is 180 × 35,980 = $6{\cdot}47$ million.

Question 2. The only knowledge we have about the signal is confined to the 1 minute interval, T, we've recorded. This means that the information carried by the observed pattern is completely indistinguishable from that we'd get from a periodic signal which repeats itself with a period, T. We can therefore apply Fourier methods to obtain a spectrum showing the amplitudes and phases at a series of frequencies, 0, f_0, $2f_0$, $3f_0$, etc, where $f_0 = 1/T$. This means we can't resolve spectral details which are closer together in frequency than an interval, f_0. In this case $T = 60$ seconds, so the resolution will be $f_0 = 1/60$th of a Hertz. (Note that this result applies because we have no 'extra' knowledge about the signal so it consists of an otherwise unpredictable pattern of frequencies. In some specific cases we may already 'know' something else about the signal. We might, for example, already have reason to know that the signal is 'really' a single sinewave. Under these conditions we can process the spectrum further to obtain a more accurate determination of its frequency, amplitude, and phase. Such a measurement would be limited only by the signal/noise ratio of the input signal. However we can't do this without the 'extra' knowledge about the signal's form. The point is that such a single sinewave signal with a frequency $f \neq nf_0$ will produce an 0, f_0, $2f_0$, $3f_0$, ..., spectrum which is indistinguishable — during the limited observation time — from one produced by some other suitable combination .)

Chapter 8

Question 1. The S/N power ratio is equal to the <u>square</u> of the ratio of the signal/noise voltages. Knowing this we use equation 8.13 to say that the data capacity is

$$C = 10000 \times \log_2\left\{1 + \left(\frac{1}{0 \cdot 001}\right)^2\right\} = 199{,}314 \text{ bits per second}$$

(N.B. this answer assumes that $\log_2\{x\} = 3{\cdot}3219 \times \log_{10}\{x\}$. If we use the easier to remember approximation of $\log_2\{x\} = 3{\cdot}33 \times \log_{10}\{x\}$ we get the less accurate result of 199,800 bps. In practice this slight difference isn't likely to lead to problems, but you should bear in mind that the accuracy of the result is affected when you use the rougher approximation.)

Question 2. The bandwidth is 100 kHz, hence the channel can carry a serial stream of 2×100,000 bps. Since each sample requires 8 bits this means the channel can carry up to 200,000/8 = 25,000 samples per second. From equation 8.21 we can say that the number of bits per sample we can get through a given channel will be such that

$$m \leqslant \frac{S}{3kTW}$$

This result arises from the requirement that the channel's data capacity must at least equal the data rate.

W is the bandwidth of the signal we're sampling, S is the signal power, T is the noise temperature, and k is Boltzmann's constant. Since the sampling rate is 25,000 per second W must equal 25,000/2 = 12,500 Hz (i.e. 1/8th of the channel bandwidth — no surprise since we're sending 8 bits per sample). Rearranging the above we discover that the noise temperature must be

$$T \leqslant \frac{S}{3kW m} = \frac{10^{-6}}{3 \times 1 \cdot 38 \times 10^{-23} \times 12500 \times 8} = 2 \cdot 4 \times 10^{11} \text{ K}$$

(Remember that this isn't the 'real' temperature of the system. It is the temperature a thermal noise source would need in order to produce the same amount of noise.)

The same argument applied to an analog signal means that we now require

$$I \leqslant C_{analog}$$

which, from expressions 8.17 and 8.15, is equivalent to saying that

$$2mW \leqslant W \log_2 \left\{ 1 + \frac{S}{kTW} \right\}$$

which can be rearranged into the inequality

$$T \leqslant \frac{S}{(2^{2m} - 1)kW}$$

Using the values provided in the question this leads to a noise temperature of 88 million K. Comparing the digital and analog results we can see that an analog transmission requires a much lower channel noise level to equal the performance of the digital system.

Chapter 9

Question 2. Since CD uses 16-bit samples it is able to indicate 2^{16} distinct voltage (or sound pressure) levels. This means that the ratio between the largest and smallest variations it can indicate will be around $(2^{16} : 1) =$ 65,536. It is conventional to express S/N ratios and dynamic ranges as <u>power</u> ratios in decibels. The above value corresponds to a power ratio of $65,536^2$ or 96·3 dB. The bandwidth can be up to half the sampling rate, i.e. up to 22·05 kHz. Taking 44,100 samples per second, for two audio channels, with 16-bit samples means we generate audio data at the rate of 1,411,200 bps. 1 hour corresponds to 3600 seconds so the total number of audio bits recorded on a 1 hour CD will be $5·08 \times 10^9$ bits.

Chapter 12

Question 1. The reference level for LP recording is a peak velocity of 5 cm/s. For a sinewave this corresponds to an rms velocity of $5 / \sqrt{2} = 3·53$ cm/s. The cartridge's sensitivity is 0·2 mV/cm/s, so it will produce an output of $3·53 \times 0·2 = 0.707$ mV rms when playing a 0 dB reference level signal. A +26 dB signal has a power $10^{26/10} = 398$ times that of a 0 dB tone. Since the voltage varies with the square root of the power we can expect the +26 dB signal to have a velocity and voltage level $\sqrt{398} = 19·95$ times greater than a 0 dB signal. Hence when playing this tone the cartridge will produce an output of $0·707 \times 19·95 = 14·1$ mV rms.

Question 2. The peak velocity of the above sinewave signal will be $5 \times 19.95 = 99.75$ cm/s. From expression 12.4 we can see that this peak velocity corresponds to the factor $2\pi f A$ where A is the peak amplitude and f is the sinewave frequency. We can therefore work out that $A = 0.0158$ cm. The recorded modulation can therefore swing over a range of $2A = 0.0317$ cm or 31,751 steps of 10 nm. Ignoring any smoothing effects produced by the stylus resting on many molecules this implies a dynamic ratio of $31751^2 = 90$ dB.

Chapter 13

Question 2. To work out the noise factor we can assume that the source produces a thermal noise spectral density of

$$e_s^2 = 4kTR = 4 \times 1.38 \times 10^{-23} \times 300 \times 22000 = 3.64 \times 10^{-16} \text{ V}^2/\text{Hz}$$

The values given in the question tell us that $e_n^2 = (5 \times 10^{-9})^2 = 2.5 \times 10^{-17}$ V²/Hz, and $i_n^2 = (10^{-12})^2 = 10^{-24}$ A/Hz. Using expression 13.10 we can therefore say that the amplifier's noise factor value when used in this situation will be

$$F = \frac{3.64 \times 10^{-16} + 2.5 \times 10^{-17} + 10^{-24} \times (22000)^2}{3.64 \times 10^{-16}} = 2.39$$

From expression 13.13 we can say that the amplifier's noise temperature value will be

$$T_n = \frac{e_n^2 + i_n^2 R_s^2}{4kR_s} = 419 \text{ K}$$

Chapter 14

Question 2. The optimum S/N will occur when we arrange for the source resistance presented to the amplifier to be equal to e_n / i_n. This means we require a resistance

$$R_s' = \frac{e_n}{i_n} = \frac{4 \times 10^{-9}}{10^{-13}} = 40,000 \text{ } \Omega$$

The actual source resistance is $R_s = 10$ kΩ. When using a turns ratio of β the transformed resistance is $R_s' = \beta^2 R_s$ so we can say that the required turns ratio must be

$$1 : \beta = 1 : \sqrt{\frac{R_s'}{R_s}} = 1 : 2$$

The best signal power transfer would occur if we arranged for a transformed source resistance which equals the amplifier's input resistance, i.e.

$$R_s' = \beta^2 R = R_{in}$$

Since $R_s = 10$ kΩ and $R_{in} = 100$ kΩ this means we require a value of $\beta = \sqrt{10} = 3 \cdot 16$, i.e. a turns ratio of 1:3·16.

Question 3. Using the transformer which provides optimum S/N performance we have a source resistance — as seen by the amplifier — of $R_s' = 40$ kΩ. The amplifier doesn't know the transformer exists. So far as it is concerned it has a 40 kΩ source connected to its input which is generating thermal noise. The noise spectral density from the source therefore appears to be

$$e_s^2 = 4kTR_s' = 6 \cdot 6 \times 10^{-16} \text{ V}^2/ \text{Hz}$$

By looking back at section 13.3 we can find expression 13.5 which tells us the output noise spectral density, E_0^2, as a function of the amplifier's noise, its gain, and the source resistance and noise level. Using the above value for source noise, recognising that the effective source resistance is 40 kΩ, and taking the amplifier noise and resistance values from question 2 we can therefore work out that $E_0 = 18 \text{ } \mu\text{V}/\sqrt{\text{Hz}}$.

Question 4. To answer this question we can use expression 14.13. This produces the total noise factor value

$$F = 1 \cdot 1 + \frac{2 \cdot 5 - 1}{10} = 1 \cdot 25$$

Chapter 15

Question 2. The time constant value equals the resistance × the capacitance. In this case this produces the value 100,000 × 0·00001 = 1. The units of Ohms and Farads are defined such that Farads × Ohms = Seconds, so the time constant in this case is 1 second.

The analog integrator's power gain value can be worked out using expression 15.8. Using the values given in the question we get

$$G \left\{ 1 \text{ kHz} \right\} = \frac{\text{Sin}^2 \left\{ \pi \times 5 \cdot 25 \times 10 \right\}}{(\pi \times 5 \cdot 25 \times 1)^2} = 0 \cdot 003676$$

This corresponds to −24 dB. The integrator therefore strongly attenuates signal fluctuations at this frequency.

Question 3. To answer this question we can use expression 15.12. This tells us the signal to noise power ratio we'll obtain from a given sequence of analog integrations in the presence of white noise. In this case we're looking for the signal power which we would be able to observe with a signal to noise ratio of unity. We can rearrange 15.12 to obtain

$$v = \sqrt{\frac{SP_S}{2ptN}}$$

where v is the input signal voltage, p is the number of integrations, t is the duration of each integration, P_S is the <u>output</u> (i.e. integrated) signal power, and N is the output signal power. S is the input noise power spectral density. Note that since this is a <u>power</u> spectral density, it should be in units of W/Hz. As is common in engineering, however, we have been given a noise level, e_n, in units of volts per root hertz. The standard relationship between voltages and powers is that $P = V^2/R$ where R is the resistance across which the voltage appears. We can therefore work out the noise power spectral density using the expression

$$S = e_n^2/R$$

As was stated at the start of chapter 15, the expressions derived in the chapter have been simplified by assuming that the load/source resistances everywhere are one Ohm. We can therefore say that $S = e_n^2 = 10^{-16}$ W/Hz. (By assuming throughout the argument that all load/source resistances are one Ohm we have, in fact, produced a set of expressions where the noise spectral densities are essentially in units of volts squared per Hertz so the above is equivalent to $S = e_n^2 = 10^{-16}$ V²/Hz.)

In this case we're considering the situation when $P_S/N = 1$ which means that the observed input signal voltage will be

$$v = \sqrt{\frac{S}{2pt}}$$

Using the values provided this leads to an input of $v = 0.7$ nanovolts.

Chapter 18

Question 3. Since there are four symbols we need at least 4 distinct patterns of bits to represent them. A fixed length representation therefore requires at least n bits where $2^n \geqslant 4$, i.e. we require $n = 2$ bits (or more).

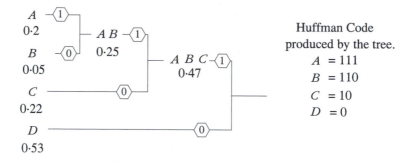

Huffman Code
produced by the tree.

$A = 111$
$B = 110$
$C = 10$
$D = 0$

The 'tree' diagram shown above can be used to generate a Huffman code for the four-symbol set described in the question. A typical message 512 symbols long would contain $0\cdot2 \times 512 = 102\cdot4$ A's, $0\cdot05 \times 512 = 25\cdot6$ B's, $0\cdot22 \times 512 = 112\cdot6$ C's, and $0\cdot53 \times 512 = 271\cdot4$ D's. Using the Huffman code we have obtained means we need to send 1 bit per D, 2 bits per C, and 3 bits per A or B. The total number of bits required for an average 512 symbol message will therefore be

$$102\cdot4 \times 3 + 25\cdot6 \times 3 + 112\cdot6 \times 2 + 271\cdot4 \times 1 = 880\cdot6.$$

(N. B. We can allow a non-integer number of bits here since we're talking about an average value.) The specific message described in the question requires $256 \times 3 = 768$ bits to transmit the A's and $256 \times 1 = 256$ bits to transmit the D's. Hence it requires a total of 1024 bits.

Chapter 20

Question 3. A 22-bit shift register can store $2^{22} = 4,194,304$ different patterns of '1's and '0's. One of these will be the 'inaccessible' state which a maximal length generator must avoid. As a result the system will cycle through 4,194,303 states before repeating itself. At a clock rate of 100,000 states/second this means that it will take $4,194,303/100,000 = 41\cdot91$ seconds before repeating itself.

Appendix 2

Programs

The following programs have been written to help you explore some of the ideas described in this book. Each program is presented in two versions: a *BBC BASIC* version and a *'C'* one. My personal preference is for simple programs is BBC BASIC. Unlike most basic 'dialects', the BBC form is well structured, powerful, and easy to use. For those who (like me!) aren't highly skilled in programming it also has the advantage of being easier to read and understand than 'C'.

The main disadvantage of BBC BASIC is that it's use is much less widespread than 'C'. This is a pity since it — and the RISC OS computers which are its main home — have many practical advantages over more common machines and languages. The BASIC programs are therefore provided for those who, like me, are not professional computer programmers, and for those fortunate enough to have access to a RISC OS computer. The 'C' versions are provided as programs which should be highly portable, although less readable by mere humans! The chances are that — whatever computer you use — the 'C' versions should run as given here whereas the BASIC versions will probably need modifying for computers that don't use RISC OS.

Chapter 4 — Getting the message

The last question at the end of chapter 4 invites you to write a computer program to discover how the number of bits transmitted correctly varies with the chosen decision level (the level used by the receiver to distinguish a '1' from a '0'). The main purpose is to show you that the best choice is, indeed, normally mid-way between the '0' logic level, V_0 and the '1' logic level, V_1. The following program can be used to answer the question and satisfy yourself on this point. If you wish you could try modifying the program and discovering what happens when the numbers of '1's and '0's are different, or the effect of a noise level (sigma) which differs for '1's and '0's. You should then find that — under these 'non-symmetric' conditions the mid-level isn't always the best choice!

'C' program showing number of bits received correctly

```
#include <stdio.h>
#include <math.h>
int n1,n0,bits;
float v1,v0,sigma,v_step;
float r2,v_decide,correct;
float erf(float);
float compute(void);


main()
{
  r2=sqrt(2.0);
  printf("Enter v0 and v1 > ");
  scanf("%f %f",&v0,&v1);
  v_step=(v1-v0)/20.0;
  printf("\nEnter sigma = ");
  scanf("%f",&sigma);
  printf("\nEnter total number of
bits > ");
  scanf("%i",&bits);
  n0=bits/2;
  n1=bits/2;
  v_decide=v0;
  printf(" v_decide   #bits ok\n");
  do
  {
    correct=compute();
    printf("%6.3f
```

```
%6.0f\n",v_decide,correct);
    v_decide+=v_step;
  } while (v_decide <= v1);
}


float compute(void)
{
  float c1,c0,answer;
  c1=r2*(v1-v_decide)/sigma;
  c0=r2*(v_decide-v0)/sigma;
  c1=0.5*(1.0+erf(c1));
  c0=0.5*(1.0+erf(c0));
  answer=n0*c0+c1*n1;
  return answer;
}


float erf(float xin)
{
  float t;
  t=1.0/(1.0+0.47*xin);
  t=1-(0.348*t-0.0958*t*t+0.748*t*t*
t)/exp(xin*xin);
  return t;
}
```

BBC BASIC program showing number of bits received correctly

```
INPUT "Input v0 and v1",v0,v1
v_step=(v1-v0)/20
INPUT "Sigma = ",sig
INPUT " enter number of bits", bits%
n0%=bits%/2
n1%=bits%/2
v_decide=v0
PRINT" v_decide    #bits ok"
WHILE v_decide<=v1
  correct%=FNcompute
  PRINT v_decide,correct%
  v_decide+=v_step
ENDWHILE
```

```
END
:
DEFFNcompute
C1=SQR(2)*(v1-v_decide)/sig
C0=SQR(2)*(v_decide-v0)/sig
C1=0.5*(1+FNerf(C1))
C0=0.5*(1+FNerf(C0))
=n0%*C0+n1%*C1
:
DEFFNerf(xin)
t=1/(1+0.47*xin)
=1-(0.348*t-0.0958*t*t+0.748*t*t*t)/
EXP(xin*xin)
```

274

Chapter 7 — Fourier Transforms

The following illustrates how we can use a Fourier Transform to compute the spectrum of a signal observed over a specific interval. The program takes an input sinewave and generates 64 data samples. These samples are then transformed to produce a spectrum of A_n and B_n values (sin and cos amplitude components) which could be used to reconstruct the waveform using an expression like 7.3. A simple 'fitting' method is used to indicate how the spectrum can be used to estimate the actual frequency of the input sinewave. Note that this fitting only means something if we already 'know' that the input <u>is</u> a portion of a single sinewave. Note also that the numerical methods used in this program are deliberately fairly simple. The program therefore lacks both elegance and accuracy! Much better methods can be found by looking in appropriate books on applied maths or computer programming. In particular you're strongly recommended to use one of the *Fast Fourier Transform* (FFT) routines listed in computing textbooks whenever you want to process more than a few data points. Various FFT routines are available, all with their own good/bad points. However, they're all much quicker than the 'slow' methods used here. Their main disadvantage is that the way they work is very difficult to understand! An example of an FFT program is included after the 'slow' example listed below.

'C' program showing the use of a Fourier Transform

```
#include <stdio.h>
#include <math.h>

float generate(void);
float fourier(void);
float fit(void);
float get_mean(void);
float get_level(int);

int points=64;
int now;
float data[65], cos_amp[65],
sin_amp[65];
float pi=3.1415927;
float f_in;
float f0,f;
float dc;
```

```
main()
{
  f0=2.0*pi/64.0;
  printf("N.B. All frequencies in
number of cycles\nduring observed
interval.\n\n");
  printf("Enter i/p frequency
(cycles) > ");
  scanf("%f",&f_in);
  generate();
  dc=get_mean();
  fourier();
  fit();
}

/* Generate creates data points from
   the frequency & phase provided.
*/
```

275

```
float generate(void)
{
  now=0;
  printf("\n\nData points are :-\n");
  do
  {
    data[now]=sin(f_in*f0*now);
    printf("%5.2f  ",data[now]);
    if ( now%8==7 ) printf("\n");
    now++;
  } while (now < 64);
}


/*  Get_mean recovers and suppresses
the d.c. level
*/


float get_mean(void)
{
  float disc;
  now=0;
  disc=0.0;
  do
  {
    disc+=data[now];
    now++;
  } while ( now < 64 );
  disc=disc/64.0;
  now=0;
  do
  {
    data[now]-=disc;
    now++;
  } while ( now < 64 );
  return disc;
}


/*  Fourier works out the spectrum in
the form
    cos_amp*COS + sin_amp*SIN
components
    and prints the results.
*/
```

```
float fourier(void)
{
  float cnow,snow;
  int fn;
  cos_amp[0]=0;
  sin_amp[0]=0;
  fn=1;
  f=f0;
  do
  {
    cnow=0.0;
    snow=0.0;
    now=0;
    do
    {
      snow+=data[now]*sin(f*now);
      cnow+=data[now]*cos(f*now);
      now++;
    } while ( now < 64 );
    cos_amp[fn]=cnow/32.0;
    sin_amp[fn]=snow/32.0;
    fn++;
    f=fn*f0;
  } while ( fn < 63 );
  fn=0;
  f=0.0;
  printf("\n\n  freq  sin_amp  cos_amp
");
  printf("  freq  sin_amp  cos_amp
");
  do
  {
    printf("%2i %6.3f %6.3f
",fn,sin_amp[fn],cos_amp[fn]);
    fn++;
    f=fn*f0;
    printf("%2i %6.3f
%6.3f\n",fn,sin_amp[fn],cos_amp[fn]);
    fn++;
    f=fn*f0;
  } while ( fn < 32 );
}


/*
   Fit locates the spectral point
having the
   greatest power and uses the levels
```

276

```
of the
  two point either side to roughly
estimate
  the frequency of the input
sinewave.
*/

float fit(void)
{
  float up,down,power,f_fitted;
  float dp,dm,a,peak;
  int now,peak_at;
  now=1;
  peak=0.0;
  do
  {
    power=get_level(now);
    if ( power > peak )
    {
      peak_at=now;
      peak=power;
    }
    now++;
  } while ( now < 33 );
  up=get_level(peak_at+1);
```

```
  down=get_level(peak_at-1);
  power=get_level(peak_at);
  dp=up-power;
  dm=power-down;
  a=2.0*dm/(dp-dm);
  f_fitted=peak_at-1.0-a;
  printf(" peak component at f = %3i
cycles.\n",peak_at);
  printf(" f fitted = %6.3f
cycles.\n",f_fitted);

}


/*  get_level provides the modulus of
the power   */

float get_level(int now_at)
{
  float answer;
  answer=cos_amp[now_at]*
cos_amp[now_at];
  answer+=sin_amp[now_at]*
sin_amp[now_at];
  return sqrt(answer);
}
```

BASIC version of Fourier demonstration program

```
REM BASIC Fourier Demonstation

points%=64
DIM
data(points%),cos_amp(points%),sin_am
p(points%)
f0=2.0*PI/points%
PRINT "N. B. All frequencies in
numbers of cycles"
PRINT "during the observed interval."
INPUT "Enter i/p frequency (cycles) >
",f_in
PROCgenerate
dc=FNget_mean
PROCfourier
PROCfit
END
```

```
:
DEFPROCgenerate
now%=0
PRINT CHR$(13)+CHR$(13)+"Data points
are :-"
REPEAT
  data(now%)=SIN(f_in*f0*now%)
  PRINT FNpoint(data(now%),2);
  IF now%MOD8 = 7 : PRINT
  now%+=1
UNTIL now%=points%
ENDPROC
:
DEFFNget_mean
disc=0.0 : now%=0
REPEAT
```

```
  disc+=data(now%)
  now%+=1
UNTIL now%=points%
disc=disc/points%
data()=data()-disc
=disc
:
DEFPROCfourier
cos_amp(0)=0 : sin_amp(0)=0
fn%=1 : f=f0
REPEAT
  cnow=0.0 : snow=0.0
  now%=0
  REPEAT
    snow+=data(now%)*SIN(f*now%)
    cnow+=data(now%)*COS(f*now%)
    now%+=1
  UNTIL now%=points%
  cos_amp(fn%)=cnow*2.0/points%
  sin_amp(fn%)=snow*2.0/points%
  fn%+=1
  f=fn%*f0
UNTIL fn% = (points%-1)
fn%=0
f=0.0
PRINT : PRINT
PRINT " freq    sin_amp    cos_amp
";
PRINT "    freq    sin_amp
cos_amp"
REPEAT
PRINT fn%;"
";FNpoint(sin_amp(fn%),3);"
";FNpoint(cos_amp(fn%),3);"    ";
fn%+=1
f=fn%*f0
PRINT "    ";fn%;"
";FNpoint(sin_amp(fn%),3);"
";FNpoint(cos_amp(fn%),3)
fn%+=1
f=fn%*f0
UNTIL fn%>=32
ENDPROC
:
DEFPROCfit
```

```
now%=1
peak=0.0
REPEAT
  power=FNget_level(now%)
  IF power>peak : peak_at%=now% :
peak=power
  now%+=1
UNTIL now%=33
up=FNget_level(peak_at%+1)
down=FNget_level(peak_at%-1)
power=FNget_level(peak_at%)
dp=up-power
dm=power-down
a=2.0*dm/(dp-dm)
f_fitted=peak_at%-1.0-a
PRINT "Peak component at f =
";peak_at%;"  cycles"
PRINT " f fitted =
";FNpoint(f_fitted,3);"  cycles"
ENDPROC
:
DEFFNget_level(inow%)
answer=cos_amp(inow%)^2+
sin_amp(inow%)^2
=SQR(answer)
:
REM  The 'point' function lets us
print out
REM  in a flexible float format &
mimics
REM  the control provided by
'printf'.
:
DEFFNpoint(pval,after%)
pthis$=STR$(pval)
com%=INSTR(pthis$,".",0)
pfront$=LEFT$(pthis$,com%+after%)
com%=INSTR(pthis$,"E",1)
IF com%>0 THEN
  pend$=RIGHT$(pthis$,LEN(pthis$)-
com%+1)
  pfront$=pfront$+pend$
ENDIF
=pfront$
:
```

Fast Fourier transformation

The following program shows a specific example of an FFT routine. The 'C' program example shows how the `fft()` routine can be used to produce the spectrum produced by a sinewave input. The BBC BASIC version just lists the `fft` and `bit_rev` procedures which do all the work. Note that the method used requires the input data to be stored in an array in the form:

x[1] = first data point; x[3]= second data point; x[5] = third data point; etc, with the even-numbered array locations all initially holding zeros.

The output spectrum appears in the form:

x[1] = sin amplitude term, 1 cycle during observation; x[2] = cos amplitude term, 1 cycle; x[3] = sin ampl. 2 cycles; etc.

As with most FFT methods, the program requires the number of data point (n or n%) to be an integer power of 2. The array must, of course, be at least twice as big as n to satisfy this requirement. The following example uses just 256 values, but the method can be used with many more points if required. When used to process many thousands of data points the FFT is likely to prove much swifter than the 'slow' transform method illustrated earlier.

'C' example of the use of an FFT routine

```
#include <stdio.h>
#include <math.h>

void fft(void);
void bit_rev(void);

int n=256;
int nn=512;
float x[514];
int here,here2,fcount;
float f,f0,pi,phase;
float rpart,ipart;


main()
{
  pi=3.1415927;
```

```
f0=2.0*pi/256;
printf("Input number cycles > ");
scanf("%f",&f);
f=f*f0;
here=0;
here2=0;
do
{
  x[here2]=sin(f*(here-1));
  here2++;
  x[here2]=0.0;
  here2++;
  here++;
} while ( here2 < 513 );
fft();
here=1;
fcount=0;
do
```

```
  {
    rpart=x[here]/128.0;
    here++;
    ipart=x[here]/128.0;
    here++;
    printf("%3i   %6.3f
%6.3f\n",fcount,rpart,ipart);
    fcount++;
  } while ( fcount < 16 );
}


void fft(void)
{
  int mmax,istep,m,i,j;
  float theta,wpi,wpr;
  float wi,wr,tr,ti,wtemp;

  mmax=2;
  theta=2.0*pi/mmax;

  bit_rev();

  while ( nn > mmax )
  {
    istep=2*mmax;
    theta=2.0*pi/mmax;
    wpr=-2.0*sin(0.5*theta)*sin(0.5*
theta);
    wpi=sin(theta);
    wr=1.0;
    wi=0.0;
    m=1;
    do
    {
      i=m;
      do
      {
        j=i+mmax;
        tr=wr*x[j]-wi*x[j+1];
        ti=wr*x[j+1]+wi*x[j];
        x[j]=x[i]-tr;
        x[j+1]=x[i+1]-ti;
```

```
        x[i]=x[i]+tr;
        x[i+1]=x[i+1]+ti;
        i+=istep;
      } while ( i < nn );
      wtemp=wr;
      wr=wr*wpr-wi*wpi+wr;
      wi=wi*wpr+wtemp*wpi+wi;
      m+=2;
    } while ( m < mmax );
    mmax=istep;
  }
}


/* bit_rev shuffles the data before
transforming   */

void bit_rev(void)
{
  int i,j,m;
  float tr,ti;
  j=1;
  i=1;
  do
  {
    if ( j > i )
    {
      tr=x[j];
      ti=x[j+1];
      x[j]=x[i];
      x[j+1]=x[i+1];
      x[i]=tr;
      x[i+1]=ti;
    }
    m=nn/2;
    while ( m >= 2 && j > m )
    {
      j-=m;
      m=m/2;
    }
    j=j+m;
    i+=2;
  } while ( i < nn );
}
```

280

The following BASIC program just consists of the procedures which are required to carry out the FFT.

BBC BASIC FFT procedures	
	NEXT
DEFPROCfft	mmax%=istep%
PROCbit_rev	ENDWHILE
mmax%=2	ENDPROC
WHILE nn%>mmax%	:
istep%=2*mmax%	DEFPROCbit_rev
theta=2*PI/mmax%	j%=1
wpr=-2*SIN(0.5*theta)^2	FOR i%=1TOnn%STEP2
wpi=SIN(theta)	IF j%>i% THEN
wr=1	tR=X(j%)
wi=0	tI=X(j%+1)
FOR m%=1TOmmax%STEP2	X(j%)=X(i%)
FOR i%=m%TOnn%STEPistep%	X(j%+1)=X(i%+1)
j%=i%+mmax%	X(i%)=tR
tr=wr*X(j%)-wi*X(j%+1)	X(i%+1)=tI
ti=wr*X(j%+1)+wi*X(j%)	ENDIF
X(j%)=X(i%)-tr	m%=nn%/2
X(j%+1)=X(i%+1)-ti	WHILE m%>=2 AND j%>m%
X(i%)=X(i%)+tr	j%=j%-m%
X(i%+1)=X(i%+1)+ti	m%=m%/2
NEXT	ENDWHILE
wtemp=wr	j%=j%+m%
wr=wr*wpr-wi*wpi+wr	NEXT
wi=wi*wpr+wtemp*wpi+wi	ENDPROC

Sinc oversampling

The following program demonstrates how it is possible to use the Sinc function to generate interpolated 'oversamples' from a set of data samples. The program only performs summation over a few data points generated from a few randomly chosen sinewave components. The resulting values are 'plotted' using a fairly crude method, chosen purely because it is likely to work on most computers. Better results can, of course, be obtained by modifying the program to increase the summing range and a graphical output which exploits the features of your particular computer.

281

'C' program showing Sinc function interpolation

```c
#include <stdio.h>
#include <stdlib.h>
#include <math.h>

int k=64;   /* number of samples  */
int r=4;    /* oversampling ratio */
float T=1.0; /* signal length      */
int c=3;    /* number components  */
int i_range=8; /* sinc calc. range */
float f0,dt,ddt;
float norm,two_pi;
float x[65]; /* sampled values     */
float fr[3]; /* freq values        */
float am[3]; /* amplitude values   */
float ph[3]; /* phase values       */
char slice[51];/* display slice     */
void generate_components(void);
void generate_signal(void);
void display(void);
float level(float);
void oversamples(int);
float sinc(float);

main()
{
  f0=1.0/T; /* scale frequency     */
  dt=T/k;/*timestep between samples*/
  ddt=dt/r; /*timestep oversamples */
  two_pi=2.0*3.1415927;
  generate_components();
  generate_signal();
  display();
  printf("\n\n   * = actual samples
o = sinc fitted oversamples\n\n");
}
void generate_components(void)
{
  int i=0;
  char dummy;
  norm=0.0;
  while ( i < c )
  {
    fr[i]=f0*k*(rand()%1000)/8000;
    ph[i]=two_pi*(rand()%1000)/1000;
    am[i]=(rand()%1000)/1000.0;
```

```c
    norm+=am[i];
    printf("f = %6.3f  a = %6.3f  ph
= %6.3f\n",fr[i],am[i],ph[i]);
    i++;
  }
  norm=1.0/norm;
  printf("\nnorm = %6.3f\n\n Press
return\n",norm);
  dummy=getchar();
}

void generate_signal(void)
{
  int i;
  float this,now;
  i=0;
  this=0.0;
  now=0.0;
  while ( i < 64 )
  {
    this=0.0;
    now=dt*i;
    x[i]=level(now);
    i++;
  }
}

float level(float time)
{
  int j;
  float answer;
  j=0;
  answer=0.0;
  while ( j < c )
  {
    answer+=am[j]*sin(two_pi*fr[j]*
time+ph[j]);
    j++;
  }
  answer=answer*norm;
  return answer;
}
void display(void)
{
  int i,j,i_stop;
```

282

```
int here;
i=0;
i_stop=64-i_range;
while ( i < 50)
{
  slice[i]=' ';
  i++;
}
i=i_range;
while ( i < i_stop )
{
  slice[25]='|';
  here=24*x[i]+25;
  printf("\n %6.3f ",x[i]);
  if (here>=0)
  {
    slice[here]='*';
    j=0;
    while ( j < 50 )
    {
      printf("%c",slice[j]);
      j++;
    }
    slice[here]=' ';
    oversamples(i);
  }
  i++;
}
}

void oversamples(int o_start)
{
  float time_now,t_then;
  float answer,x_then;
  int i_offset,h,j;
  int count=1;
  while ( count < r )
  {
    time_now=o_start*dt+count*ddt;
    answer=0.0;
```

```
    i_offset=-i_range;
    while ( i_offset < i_range)
    {
      x_then=x[o_start+i_offset];
      t_then=dt*(o_start+i_offset);
      answer+=x_then*sinc(3.1415927*
(time_now-t_then)/dt);
      i_offset++;
    }

    slice[25]='|';
    h=24*answer+25;
    printf("\n %6.3f ",answer);
    if (h>=0)
    {
      slice[h]='o';
      j=0;
      while ( j < 50 )
      {
        printf("%c",slice[j]);
        j++;
      }
      slice[h]=' ';
    }
    count++;
  }
}

float sinc(float sc)
{
  float answer;
  if ( sc != 0.0 )
  {
    answer=sin(sc)/sc;
  }
  else
  {
    answer=1.0;
  }
  return answer;
}
```

BBC BASIC version of sinc oversampling program

```
K%=64 : R%=4 : C%=3
T=1.0 :
DIM x(65),fr(3),am(3),ph(3)
DIM slice% 51
slice%?51=13
f0=1/T : dt=T/K% : ddt=dt/R%
two_pi=2*PI
i_range%=8
PROCgenerate_components
PROCgenerate_signal
PROCdisplay
PRINT : PRINT
PRINT "  * = actual samples    o =
sinc fitted oversamples "
END
:
DEFPROCgenerate_components
I%=0 : norm=0.0
REPEAT
   fr(I%)=f0*K%*RND(1)/8
   ph(I%)=two_pi*RND(1)
   am(I%)=RND(1)
   norm+=am(I%)
   PRINT "f = ";fr(I%);"  a =
";am(I%);"  ph = ";ph(I%)
   I%+=1
UNTIL I%=C%
norm=1.0/norm
PRINT : PRINT
PRINT "norm = ";norm
PRINT " press return "
wait$=GET$
ENDPROC
:
DEFPROCgenerate_signal
LOCAL I%,this,now
I%=0 : this=0.0 : now=0.0
REPEAT
  this=0.0
  now=dt*I%
  x(I%)=FNlevel(now)
  I%+=1
UNTIL I% = 64
ENDPROC
:
```

```
DEFFNlevel(time)
LOCAL J%,answer
J%=0 : answer=0.0
REPEAT
   answer+=am(J%)*SIN(two_pi*fr(J%)*
time+ph(J%))
   J%+=1
UNTIL J%=C%
answer=answer*norm
=answer
:
DEFPROCdisplay
LOCAL i%,j%,i_stop%,here%
i%=0
i_stop%=63-i_range%
REPEAT
 slice%?i%=ASC" "
 i%+=1
UNTIL i%=51
slice%?i%=13
i%=i_range%
REPEAT
  slice%?25=ASC"|"
  here%=24*x(i%)+25
  PRINT FNpoint(x(i%),3)+" ";
  IF (here%>0) THEN
    slice%?here%=ASC"*"
    PRINT $slice%
    slice%?here%=ASC" "
    PROCoversamples(i%)
  ENDIF
  i%+=1
UNTIL i%=i_stop%
ENDPROC
:
DEFPROCoversamples(o_start%)
LOCAL time_now,t_then,answer,x_then
LOCAL i_offset%,h%,j%,count%
count%=1
REPEAT
   time_now=o_start%*dt+count%*ddt
   answer=0
   i_offset%=-i_range%
   REPEAT
     x_then=x(o_start%+i_offset%)
```

```
   t_then=dt*(o_start%+i_offset%)
    answer+=x_then*FNsinc(PI*
(time_now-t_then)/dt)
     i_offset%+=1
  UNTIL i_offset%=i_range%
  slice%?25=ASC"|"
  h%=24*answer+25
  PRINT FNpoint(answer,3)+" ";
  IF (h%>=0) AND (h%<51) THEN
    slice%?h%=ASC"o"
    PRINT $slice%
    slice%?h%=ASC" "
  ELSE
    PRINT "out range answer =
";answer;"    h = ";h%
  ENDIF
  count%+=1
UNTIL count%=R%
ENDPROC
:
DEFFNsinc(input)
```

```
LOCAL answer
IF input=0 THEN
  answer=1
  ELSE
  answer=SIN(input)/input
ENDIF
=answer
:
DEFFNpoint(pval,after%)
pthis$=STR$(pval)
com%=INSTR(pthis$,".",0)
pfront$=LEFT$(pthis$,com%+after%)
com%=INSTR(pthis$,"E",1)
IF com%>0 THEN
  pend$=RIGHT$(pthis$,LEN(pthis$)-
com%+1)
  pfront$=LEFT$(pfront$,2)
  pfront$=pfront$+pend$
ENDIF
IF LEFT$(pfront$,1)<>"-" : pfront$="
"+pfront$
=pfront$
```

Chapter 21 — Encrypting information

The following program is a simple example of an encryption and de-encryption process based on the *RSA* method described in Chapter 21. Note that, as shown below, the encryption isn't very good since the input text is only grouped into pairs before encryption. As a result, assuming that text consists of the standard English set of less than 100 characters (a, b, c, ... plus A, B, C, plus normal punctuation) there are less than 10,000 legal character pairs. Hence an encoded message a few thousand characters long would be vulnerable to entropic attack. A practical system would gather the characters into larger groups before encrypting them to avoid this problem. However, this requires the program to cope with very large integer values and necessitates more involved methods to achieve the same basic result. This is because of the finite precision with which digital computers store and processes integer values. For the same reason the program shown below only works correctly on most machines when given primes whose values are less than 255.

'C' encryption program

```
#include <stdio.h>
#include <math.h>

int n,p,q,s,t,r,e;
int here, count, ok;
char in_text[256];
int numbers[128];

int text_to_numbers(void);
int show_numbers(void);
int choose_primes(void);
int encrypt(void);
int decrypt(void);
int numbers_back_to_text(void);

main()
{
  printf("Input line of text to
encode >");
  gets(&in_text);
  count=text_to_numbers();
  printf("Count = %i\n",count);
  printf("Text converted to integers
>\n");
  show_numbers();
  ok=0;
  while (ok==0)
  {
  ok=choose_primes();
  encrypt();
  printf("\n\nEncrypted numbers >
\n");
  show_numbers();
  decrypt();
  printf("\n\nDecrypted numbers >
\n");
  show_numbers();
  numbers_back_to_text();
  }
}



/*   The proceedure converts the text
characters into a series of integers
in the range 1-99, pairs them, and
stores the resulting values.    */
```

```
int text_to_numbers(void)
{
  int first,second,so_far;
  here=0;
  so_far=0;
  do
  {
    first=in_text[here];
    second=in_text[here+1];
    if (first > 31 ) first=first-30;
    if (second > 31 ) second=second-
30;
    numbers[so_far]=first+100*second;
    so_far++;
    here=here+2;
  } while( first!=0 && second!=0);
  return so_far;
}


/*  The following prints out the
current set of numbers.*/

int show_numbers(void)
{
  int so_far, this_number;
  so_far=0;
  do
  {
    this_number=numbers[so_far];
    printf("%i  ",this_number);
    so_far++;
  } while (so_far < count);
}


/*  The following reads in the chosen
prime values.*/

int choose_primes(void)
{
  int answer;
  printf("Input a pair of primes \n
(Suggest 107 103) > ");
  scanf("%i %i",&p,&q);
  n=p*q;
```

286

```
s=(p-1)*(q-1);
printf("\nModulus  n = %i\n",n);
printf("Input a new value prime
w.r.t %i and %i \n(decode key
value).\n",p,q);
printf(" (Suggest 101) > ");
scanf("%i",&r);
e=0;
do
{
  e++;
  t=(e*r)%s;
} while ( t!=1 && e<=s );
if (t==1)
{
  answer=1;
  printf("\n OK, the encode key is;
e = %i\n",e);
}
else
{
  answer=0;
  printf("\n NO key found !\n");
}
return answer;
}
```

```
/*  The following encrypts the
information using the public key
values of the modulus, n, and the
value, e.*/
```

```
int encrypt(void)
{
  int so_far=0;
  unsigned long int
  temp,y,times;
  do
  {
    temp=numbers[so_far];
    y=temp;
    times=1;
    do
    {
      y=(y*temp)%n;
      times++;
```

```
  } while (times < e );
    numbers[so_far]=y;
    so_far++;
  } while ( so_far < count );
}
```

```
/*  This de-encrypts the encyphered
numbers by using the public modulus,
n, and the SECRET value, r.*/
```

```
int decrypt(void)
{
  int so_far=0;
  unsigned long int temp,x,times;
  do
  {
    temp=numbers[so_far];
    x=temp;
    times=1;
    do
    {
      x=(x*temp)%n;
      times++;
    } while ( times < r );
    numbers[so_far]=x;
    so_far++;
  } while ( so_far < count );
}
```

```
/*  The following turns the recovered
numbers back into text.*/
```

```
int numbers_back_to_text(void)
{
  int so_far=0;
  int first,second,this;
  printf("\n Recovered text > \n");
    do
    {
      this=numbers[so_far];
      first=this%100;
      second=this/100;
      if (first > 0)
      {
        first=first+30;
        printf("%c",first);
```

287

```
            }                                    printf("%c",second);
  if (second> 0)                              }
                                             so_far++;
  {                                        } while (so_far < count);
      second=second+30;                 }
```

BBC BASIC encryption program

```
DIM in_text% 256, numbers%(256)      stop%=FALSE
FOR I%=0 TO 256                      REPEAT
  in_text%?I%=0:numbers%(I%)=0         first%=in_text%?here%
NEXT                                   second%=in_text%?(here%+1)
:                                      IF (first%=13) THEN
REM **** Now the main routine            first%=0  : stop%=TRUE
:                                      ENDIF
PRINT "Input line of text to encode >  IF (second%=13) THEN
"                                        second%=0 : stop%=TRUE
INPUTLINE $in_text%                    ENDIF
count%=FNtext_to_numbers               IF (first%>31) : first%-=30
PRINT "Count = ";count%                IF (second%>31): second%-=30
PRINT "Text converted to integers >"   numbers%(so_far%)=first%+100*
PRINT                                second%
PROCshow_numbers                       so_far%+=1
ok%=0                                  here%+=2
WHILE (ok%=0)                        UNTIL stop% OR (here%>250)
  ok%=FNchoose_primes                =so_far%
  PROCencrypt                        :
  PRINT CHR$(13)+CHR$(13)+"Encrypted DEFPROCshow_numbers
numbers >"+CHR$(13)                  LOCAL so_far%
  PROCshow_numbers                   so_far%=0
  PROCdecrypt                        REPEAT
  PRINT CHR$(13)+CHR$(13)+"Decrypted   PRINT numbers%(so_far%);"  ";
numbers >"+CHR$(13)                    so_far%+=1
  PROCshow_numbers                   UNTIL so_far%=count%
  PROCnumbers_back_to_text           ENDPROC
ENDWHILE                             :
END                                  DEFFNchoose_primes
:                                    LOCAL answer%
REM **** Routines named and used like PRINT "Input a pair of primes"
'C' version                          INPUT "(Suggest 107,103) > ",p%,q%
:                                    n%=p%*q%
DEFFNtext_to_numbers                 s%=(p%-1)*(q%-1)
LOCAL first%,second%                 PRINT "Modulus n = ";n%
LOCAL so_far%,stop%                  PRINT "Input a new prime w.r.t.
here%=0 : so_far%=0                   ";p%;" and ";q%
```

```
PRINT "(Decode key value) "
INPUT "(Suggest 101) > ",r%
e%=0
REPEAT
  e%+=1
  t%=(e%*r%)MODs%
UNTIL (t%=1) OR (e%>=s%)
IF t%=1 THEN
  answer%=1
  PRINT "OK, the decode key is;  e =
";e%
  ELSE
  answer%=0
  PRINT "NO key found !"
ENDIF
=answer%
:
DEFPROCencrypt
LOCAL so_far%,temp%,y%,times%
so_far%=0
REPEAT
  temp%=numbers%(so_far%)
  y%=temp%
  times%=1
  REPEAT
    y%=(y%*temp%)MODn%
    times%+=1
  UNTIL times%=e%
  numbers%(so_far%)=y%
  so_far%+=1
UNTIL so_far%=count%
ENDPROC
:
DEFPROCdecrypt
LOCAL so_far%,temp%,x%,times%
```

```
so_far%=0
REPEAT
  temp%=numbers%(so_far%)
  x%=temp%
  times%=1
  REPEAT
    x%=(x%*temp%)MODn%
    times%+=1
  UNTIL times%=r%
  numbers%(so_far%)=x%
  so_far%+=1
UNTIL so_far%=count%
ENDPROC
:
DEFPROCnumbers_back_to_text
LOCAL so_far%,first%,second%,this%
so_far%=0
PRINT CHR$(13)+"Recovered text >"+
CHR$(13)
REPEAT
  this%=numbers%(so_far%)
  first%=this%MOD100
  second%=this%DIV100
  IF (first%>0) AND (first%<>13) THEN
    first%+=30
    PRINT CHR$(first%);
  ENDIF
  IF (second%>0) AND (second%<>13)
THEN
    second%+=30
    PRINT CHR$(second%);
  ENDIF
  so_far%+=1
UNTIL so_far%=count%
ENDPROC
```

Finding prime numbers

In order to use the above programs you need to select a suitable trio of prime numbers. The following program provides a simple way to locate prime integers in a given range. The program asks for the maximum and minimum values which limit the range to search. It then prints out all the primes it finds in the given range.

289

'C' prime number lister.

```c
/* finds primes in given range */

#include <stdio.h>
#include <math.h>

int min,max,now;
int ia,ib,ic, iremain;
float x,y,z, stop_at;

main()
{
  printf("Input max and min integers
>");
  scanf("%i %i",&max,&min);
  if (min>=max)
  {
    now=max;
    max=min;
    min=now;
  }
  printf("Max = %i\n",max);
  printf("Min = %i\n",min);
  now=min;
  for (now=min; now<max; now++)
  {
    stop_at=sqrt(now);
    ia=1;
    do
    {
      ia++;
      iremain=now%ia;
    } while (ia<stop_at && iremain!=
0);
    if (iremain!=0) printf(" %i PRIME
***\n",now);
  }
}
```

BBC BASIC prime finding program

```basic
REM Finds primes
:
PRINT "Input max and min integers >
";
INPUT min%,max%
IF min%>max% : SWAP min%,max%
PRINT "Max = ";max%
PRINT "Min = ";min%
now%=min%
REPEAT
  stop_at=SQR(now%)
  ia%=1
  REPEAT
    ia%+=1
    iremain%=now%MODia%
  UNTIL (iremain%=0) OR (ia%>stop_at)
  IF iremain%<>0 : PRINT now%;" is
PRIME ***"
  now%+=1
UNTIL now%>max%
END
```

Index

291